D0867362

FOUNDATIONS OF MATHEMATICAL PROGRAMMING

FOUNDATIONS OF
MATHEMATICAL
PROGRAMMING

WILLIAM W. CLAYCOMBE
University of Tennessee
Knoxville, Tennessee

WILLIAM G. SULLIVAN
University of Tennessee
Knoxville, Tennessee

RESTON PUBLISHING COMPANY, INC.
Reston, Virginia 22090
A Prentice-Hall Company

WILLIAM MADISON RANDALL LIBRARY UNC AT WILMINGTON

Library of Congress Cataloging in Publication Data

Claycombe, W   W   1943-

   Foundations of mathematical programming

    1. Programming (Mathematics) I. Sullivan,
William G., 1942-   joint author. II. Title.
QA402.5.C54   519.7   75-2050
ISBN 0-87909-282-3

© 1975 by
Reston Publishing Company, Inc.
A Prentice-Hall Company
Reston, Virginia 22090

All rights reserved. No part of
this book may be reproduced in any
way, or by any means, without
permission in writing from the
publisher.

10 9 8 7 6 5 4 3 2 1

Printed in the United States of America.

QA402
.5
.C54

Anyone who proposes to do good must not expect people to roll stones out of his way, but must accept his lot calmly even if they roll a few more on it.

Albert Schweitzer

156000

CONTENTS

PREFACE

Mathematical programming is a collection of procedures that may be used to determine optimal solutions to real-world resource allocation problems. The basic procedures are equally applicable to such diverse fields as marketing, engineering, finance, home economics and agriculture. When they are familiar with the basic procedures, practitioners can quickly recognize many problems in their field of specialization that can be formulated and solved with mathematical models.

The goal of mathematical programming is to maximize or minimize the response of a system, such as profit, chemical yield or labor cost. This response may be the function of one or more decision variables—advertising expenditures in the different media, percent of different catalysts used in a chemical reaction, product mix and so forth. The equation that expresses system response as a function of the decision variables is referred to as the objective function. When there are limitations on resources required to implement a system, they will be expressed as constraint equations. By using a mathematical model the decision maker can locate value(s) of the decision variable(s) that will result in the "best" (optimum) system in view of limited resources available.

The bulk of the effort in optimization is usually expended in gathering data for use with the model and in investigating simplifying assumptions. The actual application of the procedures is not tedious, especially with the availability of computer programs. A knowledge of the mechanics of the procedures included in this text is essential to their application, for an analyst cannot merely supply data to a computer program. Such practice frequently leads to catastrophic results.

After obtaining a mathematical solution the practitioner should validate the model. Conditions predicted by the model should be checked for their "reasonableness" with the real-world situation. The analyst's judgment, experience and mathematical sophistication must all be utilized fully at this stage of optimization.

One of the most common causes of failure with mathematical programming is a lack of validation. This occurs frequently when the analyst does not understand the real-world situation being modeled, and the individuals with practical experience are unfamiliar with the mathematical procedures. Mathematical programming is most successful when the analyst has expertise both in the field of application and with the mathematical procedures.

Mathematical programming should not be an esoteric, academic exercise. The authors have attempted to hold general philosophical discussion to a minimum and to include the largest possible number of practical examples to bring mathematical programming closer to the practitioner. Hence, we hope to assist students in developing realistic, valid models of real-world phenomena.

One criticism of mathematical programming deals with the amount of time required to gather data, obtain a solution and validate the model. Many problems require an immediate decision. The authors will agree that this is occasionally a valid criticism. However, one benefit frequently overlooked by critics is that mathematical programming gives an individual a unique sophistication and expertise in optimizing real-world situations. In many cases, when an analyst recognizes that a particular problem can be described by a certain mathematical model, the solution is apparent. Few (or no) calculations are required in these instances.

Chapter 1 is a brief review of the topics of matrix algebra relevant to the remainder of the text. No previous knowledge of matrix algebra is assumed. Chapter 2, Classical Optimization, is especially important to scientists and engineers. A basic knowledge of calculus is assumed in this chapter. If the reader does not have a background in elementary calculus, this chapter may be omitted without affecting the reader's understanding of the remainder of the text. In Chapters 3, 4 and 5, systems of linear equations are anlayzed and linear programming is treated in detail. These three chapters should be treated in sequence. Chapter 6 presents procedures that may be used to solve special types of linear programming problems with less effort than the conventional Simplex procedure. Chapter 7, Search Procedures, deals with the problem of an unknown, or intractable, objective function. This is a relatively common problem that usually is ignored in undergraduate texts.

An article entitled "How They're Planning OR at the Top" is included in the Appendix to illustrate the organization and use of operations research in United States corporations. The reader should observe from Figure 6 of this article that mathematical programming techniques constitute an important subset of the field of operations research.

We would like to express appreciation to our students for their many suggestions, and especially to Moon-Sang Lee and Farhad Momtaz for their help with the solutions manual. We would also like to thank our wives for their support; our department head, Dan Doulet, for his encouragement; and our typists, Teresa Gilbert and Sylvia Ross, for their assistance.

W. W. Claycombe
W. G. Sullivan

MATRIX ALGEBRA

INTRODUCTION

This chapter provides a presentation of the concepts in matrix algebra that are relevant to the remainder of the text. No previous knowledge of matrix algebra is assumed. An understanding of this material should provide a basic background sufficient to understand the fundamentals of mathematical programming as presented in this text. Basic Definitions, Matrix Addition and Subtraction, and Matrix Multiplication are necessary for all subsequent chapters. The Transpose of a Matrix, Identity Matrix, Determinant of a Matrix, and the Inverse Matrix form a basis for Chapters 3, 4, 5 and 6 on systems of linear equations and linear programming. Finally, eigenvalues and Hessian matrices are necessary for Chapter 2, Classical Optimization.

The reader will realize from the following presentation that matrix algebra is simply a convenient means of organizing and referencing quantitative information. This is especially important with the use of computers to store and manipulate data. Many basic concepts are easily illustrated with realistic examples. Other concepts, such as the inverse matrix and Hessian matrix, are problem-solving tools that will be explained in this chapter, and then will be used to solve realistic example problems in later chapters.

BASIC DEFINITIONS

A scalar is a single number.

Example:

$$a_1 = 2 \qquad a_2 = 1 \qquad a_3 = 3$$

A vector is a row or column of numbers.

Example:

$$A_m = \begin{bmatrix} a_1 \\ a_2 \\ \cdot \\ \cdot \\ \cdot \\ a_m \end{bmatrix} \qquad A_4 = \begin{bmatrix} 1 \\ 4 \\ 2 \\ 11 \end{bmatrix}$$

$$B_n = [b_1 \; b_2 - - - b_n]$$

$$B_3 = [6 \; 2 \; 14]$$

A subscript, m or n in this example, may be used to denote the dimension of the vector. The dimension refers to how many scalars are included in the vector. This subscript may be included for convenience, or omitted. In mathematical notation vectors are enclosed in brackets as shown above.

A matrix is a rectangular array of scalars. These scalars may be referred to as the elements of the matrix.

Example:

$$A_{m,n} = \begin{bmatrix} a_{11} & a_{12} & - & - & - & a_{1n} \\ a_{21} & a_{22} & - & - & - & a_{2n} \\ \cdot & \cdot & & \cdot & \cdot & \cdot \\ \cdot & \cdot & & \cdot & \cdot & \cdot \\ \cdot & \cdot & & \cdot & \cdot & \cdot \\ a_{m1} & a_{m2} & & \cdot & \cdot & a_{mn} \end{bmatrix}$$

$$A_{3,2} = \begin{bmatrix} 4 & 16 \\ 7 & 8 \\ 11 & 1 \end{bmatrix}$$

Throughout this book, m denotes the number of rows (or equations) and n the number of columns (or unknowns). These subscripts may be included for convenience, or omitted. In mathematical notation matrices are enclosed in brackets as shown above.

A vector is a matrix in which either n or m equals 1. Also, if the number of rows and columns are equal, the matrix is a square matrix.

MATRIX

ADDITION AND

SUBTRACTION

Matrices may be added or subtracted by adding or subtracting their corresponding elements. This is possible only if the matrices are conformable. Matrices that are conformable have the same number of rows and columns.

Example:

A chemical company has four products and three plants producing these products. Various production data are recorded in matrix form below.

CURRENT INVENTORY, THOUSANDS OF POUNDS = $A_{3,4}$

		Product Number			
		1	2	3	4
	1	6	2	0	4
Plant	2	4	3	7	3
	3	1	1	1	5

CURRENT MONTH'S PRODUCTION, THOUSANDS OF POUNDS = $B_{3,4}$

		Product Number			
		1	2	3	4
	1	10	9	7	6
Plant	2	0	8	4	11
	3	7	3	2	1

CURRENT MONTH'S SALES, THOUSANDS OF POUNDS = $C_{3,4}$

Product Number

		1	3	3	4
	1	12	7	6	6
Plant	2	2	5	4	9
	3	5	3	3	4

The new inventory level, Z, is given by $A + B - C$. As a first step in finding Z, let $D = A + B$:

$$D = \begin{bmatrix} 6 & 2 & 0 & 4 \\ 4 & 3 & 7 & 3 \\ 1 & 1 & 1 & 5 \end{bmatrix} + \begin{bmatrix} 10 & 9 & 7 & 6 \\ 0 & 8 & 4 & 11 \\ 7 & 3 & 2 & 1 \end{bmatrix}$$

$$D = \begin{bmatrix} 16 & 11 & 7 & 10 \\ 4 & 11 & 11 & 14 \\ 8 & 4 & 3 & 6 \end{bmatrix}$$

Then

$$Z = D - C = \begin{bmatrix} 16 & 11 & 7 & 10 \\ 4 & 11 & 11 & 14 \\ 8 & 4 & 3 & 6 \end{bmatrix} - \begin{bmatrix} 12 & 7 & 6 & 6 \\ 2 & 5 & 4 & 9 \\ 5 & 3 & 3 & 4 \end{bmatrix}$$

and finally Z is:

NEW INVENTORY LEVEL, THOUSANDS OF POUNDS = $Z_{3,4}$

Product Number

		1	2	3	4
	1	4	4	1	4
Plant	2	2	6	7	5
	3	3	1	0	2

MATRIX

MULTIPLICATION

Two matrices $A_{m,q}$ and $B_{p,n}$ are conformable to multiplication if q = p. The product C = AB is an m by n matrix in which each element c_{ij} of C is obtained by multiplying elements of the ith row of A by the corresponding elements of the jth column of B and adding the products. (Note: $1 \leqslant i \leqslant m$ and $1 \leqslant j \leqslant n$.)

Example:

The chemical products in the previous example may be sold to wholesalers or directly to consumers. The prices are recorded in matrix form below. Denote this matrix by P.

PRICE MATRIX, $/LB. $= P_{4,2}$

		Wholesale	Retail
	1	4	5
Product	2	3	3
	3	5	6
	4	6	8

The dollar value of our current inventories at each plant is given by Z·P. The product of the Z matrix and the P matrix is a new matrix, V, shown below. Multiplication is possible because the number of columns in Z equals the number of rows in P.

$$V_{3,2} = Z_{3,4} \cdot P_{4,2} = \begin{bmatrix} 4 & 4 & 1 & 4 \\ 2 & 6 & 7 & 5 \\ 3 & 1 & 0 & 2 \end{bmatrix} \begin{bmatrix} 4 & 5 \\ 3 & 3 \\ 5 & 6 \\ 6 & 8 \end{bmatrix}$$

$$V_{3,2} = \begin{bmatrix} (4 \times 4 + 4 \times 3 + 1 \times 5 + 4 \times 6) & (4 \times 5 + 4 \times 3 + 1 \times 6 + 4 \times 8) \\ (2 \times 4 + 6 \times 3 + 7 \times 5 + 5 \times 6) & (2 \times 5 + 6 \times 3 + 7 \times 6 + 5 \times 8) \\ (3 \times 4 + 1 \times 3 + 0 \times 5 + 2 \times 6) & (3 \times 5 + 1 \times 3 + 0 \times 6 + 2 \times 8) \end{bmatrix}$$

$$= \begin{bmatrix} 57 & 70 \\ 91 & 110 \\ 27 & 34 \end{bmatrix}$$

The V matrix is summarized as follows.

VALUE OF INVENTORY, THOUSANDS OF $ = $V_{3,2}$

		Wholesale	Retail
	1	57	70
Plant	2	91	110
	3	27	34

If the dollar value of each product is given by the following vector, W,

VALUE, $/LB. = $W_{4,1}$

	1	5
	2	3
Product	3	5
	4	6

then the value of our inventory at each plant is determined in this manner:

$$Z_{3,4} \cdot W_{4,1} = \begin{bmatrix} 4 & 4 & 1 & 4 \\ 2 & 6 & 7 & 5 \\ 3 & 1 & 0 & 2 \end{bmatrix} \begin{bmatrix} 5 \\ 3 \\ 5 \\ 6 \end{bmatrix}$$

$$ZW = \begin{bmatrix} 4 \times 5 + 4 \times 3 + 1 \times 5 + 4 \times 6 \\ 2 \times 5 + 6 \times 3 + 7 \times 5 + 5 \times 6 \\ 3 \times 5 + 1 \times 3 + 0 \times 5 + 2 \times 6 \end{bmatrix} = \begin{bmatrix} 61 \\ 93 \\ 30 \end{bmatrix}$$

TOTAL INVENTORY VALUE, THOUSANDS OF $

$$\text{Plant} \begin{array}{c} 1 \\ 2 \\ 3 \end{array} \begin{bmatrix} 61 \\ 93 \\ 30 \end{bmatrix}$$

Matrices that are conformable are not commutative with respect to multiplication, that is AB ≠ BA. However, they are associative with respect to multiplication: (AB)C = A(BC). The distributive property will also hold: A(B+C) = AB + AC. Conformable matrices are commutative and associative with respect to addition: A+B = B+A and (A+B)+C = A+(B+C).

To multiply a scalar times a matrix, multiply each element of the matrix by the scalar.

Example:

From the previous example if we lose 0.10 of our current inventory as spoilage, etc. the losses for the current month will be bZ, where b = 0.10:

$$bZ = Zb = 0.10 \begin{bmatrix} 4 & 4 & 1 & 4 \\ 2 & 6 & 7 & 5 \\ 3 & 1 & 0 & 2 \end{bmatrix}$$

These calculations are summarized below.

INVENTORY LOSSES, THOUSANDS OF POUNDS

		Product Number			
		1	2	3	4
	1	0.4	0.4	0.1	0.4
Plant	2	0.2	0.6	0.7	0.5
	3	0.3	0.1	0	0.2

A set of equations can be expressed in matrix form.

If

$$a_{11}x_1 \quad + \quad a_{12}x_2 \quad + \quad --- \quad + \quad a_{1n}x_n \quad = \quad b_1$$

$$a_{21}x_1 \quad + \quad a_{22}x_2 \quad + \quad --- \quad + \quad a_{2n}x_n \quad = \quad b_2$$

$$\cdot \qquad\qquad \cdot \qquad\qquad --- \qquad\qquad \cdot \qquad\qquad \cdot$$

$$\cdot \qquad\qquad \cdot \qquad\qquad --- \qquad\qquad \cdot \qquad\qquad \cdot$$

$$\cdot \qquad\qquad \cdot \qquad\qquad --- \qquad\qquad \cdot \qquad\qquad \cdot$$

$$a_{m1}x_1 \quad + \quad a_{m2}x_2 \quad + \quad --- \quad + \quad a_{mn}x_n \quad = \quad b_m$$

then

$$A = \begin{bmatrix} a_{11} & a_{12} & --- & a_{1n} \\ a_{21} & a_{22} & --- & a_{2n} \\ \cdot & \cdot & --- & \cdot \\ \cdot & \cdot & --- & \cdot \\ \cdot & \cdot & --- & \cdot \\ a_{m1} & a_{m2} & --- & a_{mn} \end{bmatrix}$$

$$X = \begin{bmatrix} x_1 \\ x_2 \\ \cdot \\ \cdot \\ \cdot \\ x_n \end{bmatrix} \qquad B = \begin{bmatrix} b_1 \\ b_2 \\ \cdot \\ \cdot \\ \cdot \\ b_m \end{bmatrix}$$

and

$$AX = B$$

The actual multiplication of the matrices as indicated will give the set of equations shown above.

TRANSPOSE OF

A MATRIX

The transpose of matrix A is denoted by A^T or A'. To obtain the transpose of a matrix, interchange its rows and columns. That is, the first row of a matrix becomes the first column in its transpose, etc.

Example:

$$A = \begin{bmatrix} 1 & 6 \\ 0 & 11 \\ 1 & -4 \end{bmatrix}$$

$$A^T = A' = \begin{bmatrix} 1 & 0 & 1 \\ 6 & 11 & -4 \end{bmatrix}$$

$$B = \begin{bmatrix} 1 & 4 & 7 \end{bmatrix}$$

$$B^T = B' = \begin{bmatrix} 1 \\ 4 \\ 7 \end{bmatrix}$$

IDENTITY MATRIX

The identity matrix is one in which all diagonal elements are 1 (one) and all off diagonal elements are 0 (zero) as follows:

$$I = \begin{bmatrix} 1 & 0 & 0 & --- & 0 \\ 0 & 1 & 0 & --- & 0 \\ 0 & 0 & 1 & --- & 0 \\ \cdot & \cdot & \cdot & --- & \cdot \\ \cdot & \cdot & \cdot & --- & \cdot \\ \cdot & \cdot & \cdot & --- & \cdot \\ 0 & 0 & 0 & --- & 1 \end{bmatrix}$$

For any matrix A, IA = A, or AI = A. Multiplying a matrix by I is analogous to multiplying a scalar by 1.

Example:

$$\begin{bmatrix} 1 & 0 & 0 \\ 0 & 1 & 0 \\ 0 & 0 & 1 \end{bmatrix} \begin{bmatrix} 1 & 0 & 2 \\ 3 & 6 & 1 \\ 2 & 4 & 1 \end{bmatrix} = \begin{bmatrix} 1 & 0 & 2 \\ 3 & 6 & 1 \\ 2 & 4 & 1 \end{bmatrix}$$

DETERMINANT
OF A MATRIX

The determinant of a matrix A is denoted by det A or $|A|$. For a 1×1 matrix, the determinant is simply the element.

Example:

$$\text{If } B_{1,1} = [4]$$
$$\det B = |B| = 4$$

To evaluate the determinant of a larger matrix A, we use a cofactor expansion, which can be performed as follows:

1. Choose any row or column in A for the expansion.

2. Calculate the cofactor of every element a_{ij} in the chosen row or column. The cofactor of an element a_{ij} is defined as $(-1)^{i+j}$ times the determinant of the matrix formed by omitting the ith row and j^{th} column from A.

3. The determinant of A is the sum of the products of the a_{ij}'s and their cofactors.

Example:

$$\text{If } A = \begin{bmatrix} a_{11} & a_{12} \\ a_{21} & a_{22} \end{bmatrix}$$

and we choose the first row for expansion, then

$$\text{cofactor } a_{11} = (-1)^{1+1} |a_{22}| = a_{22}$$
$$\text{cofactor } a_{12} = (-1)^{1+2} |a_{21}| = -a_{21}$$

and

$$|A| = a_{11}a_{22} - a_{12}a_{21}$$

Example:

$$\text{If A} = \begin{bmatrix} 1 & 3 & 0 \\ 1 & 4 & 1 \\ 2 & 0 & 0 \end{bmatrix}$$

and we choose the first row for expansion:

$$\text{cofactor } a_{11} = (-1)^{1+1} \begin{vmatrix} 4 & 1 \\ 0 & 0 \end{vmatrix} = (+1)(0) = 0$$

$$\text{cofactor } a_{12} = (-1)^{1+2} \begin{vmatrix} 1 & 1 \\ 2 & 0 \end{vmatrix} = (-1)(-2) = 2$$

$$\text{cofactor } a_{13} = (-1)^{1+3} \begin{vmatrix} 1 & 4 \\ 2 & 0 \end{vmatrix} = (+1)(-8) = -8$$

then

$$|A| = (1)(0) + (3)(2) + (0)(-8) = 6$$

Determinants will be used in this book in solving sets of linear equations and in solving some calculus problems. Practical examples will be given later for these examples.

INVERSE OF
A MATRIX

Every square matrix, $A_{n,n}$, that has a nonzero determinant will have a unique inverse, A^{-1}, where

$$A^{-1}A = AA^{-1} = I$$

To calculate the inverse we use the cofactor matrix, C, where C_{ij} is the cofactor of element a_{ij} as defined in the previous section. This will be illustrated with a numerical example.

$$A = \begin{bmatrix} 1 & 2 & 3 \\ 3 & 2 & 1 \\ 1 & 1 & 2 \end{bmatrix} \qquad |A| = -4$$

The cofactor matrix of A is

$$C = \begin{bmatrix} (-1)^{1+1} \begin{vmatrix} 2 & 1 \\ 1 & 2 \end{vmatrix} & (-1)^{1+2} \begin{vmatrix} 3 & 1 \\ 1 & 2 \end{vmatrix} & (-1)^{1+3} \begin{vmatrix} 3 & 2 \\ 1 & 1 \end{vmatrix} \\[2ex] (-1)^{2+1} \begin{vmatrix} 2 & 3 \\ 1 & 2 \end{vmatrix} & (-1)^{2+2} \begin{vmatrix} 1 & 3 \\ 1 & 2 \end{vmatrix} & (-1)^{2+3} \begin{vmatrix} 1 & 2 \\ 1 & 1 \end{vmatrix} \\[2ex] (-1)^{3+1} \begin{vmatrix} 2 & 3 \\ 2 & 1 \end{vmatrix} & (-1)^{3+2} \begin{vmatrix} 1 & 3 \\ 3 & 1 \end{vmatrix} & (-1)^{3+3} \begin{vmatrix} 1 & 2 \\ 3 & 2 \end{vmatrix} \end{bmatrix}$$

$$C = \begin{bmatrix} 3 & -5 & 1 \\ -1 & -1 & 1 \\ -4 & 8 & -4 \end{bmatrix}$$

The adjoint of A is the transpose of the cofactor matrix, C^T.

$$\text{adj } A = \begin{bmatrix} 3 & -1 & -4 \\ -5 & -1 & 8 \\ 1 & 1 & -4 \end{bmatrix}$$

The inverse of A is the adjoint divided by its determinant.

$$A^{-1} = \frac{\text{adj } A}{|A|} = \frac{\begin{bmatrix} 3 & -1 & -4 \\ -5 & -1 & 8 \\ 1 & 1 & -4 \end{bmatrix}}{-4} = \begin{bmatrix} -3/4 & 1/4 & 1 \\ 5/4 & 1/4 & -2 \\ -1/4 & -1/4 & 1 \end{bmatrix}$$

To check our calculations, we need to show that $A^{-1} A = I$:

$$A^{-1} A = \begin{bmatrix} -3/4 & 1/4 & 1 \\ 5/4 & 1/4 & -2 \\ -1/4 & -1/4 & 1 \end{bmatrix} \begin{bmatrix} 1 & 2 & 3 \\ 3 & 2 & 1 \\ 1 & 1 & 2 \end{bmatrix} = \begin{bmatrix} 1 & 0 & 0 \\ 0 & 1 & 0 \\ 0 & 0 & 1 \end{bmatrix}$$

The inverse matrix will be used in Chapter 3 to solve sets of linear equations.

EIGENVALUES

AND

EIGENVECTORS

Let $A_{nn} \cdot X_n = \lambda X_n$, i.e.

$$\begin{bmatrix} a_{11} & a_{12} & --- & a_{1n} \\ a_{21} & a_{22} & --- & a_{2n} \\ \cdot & \cdot & --- & \cdot \\ \cdot & \cdot & --- & \cdot \\ \cdot & \cdot & --- & \cdot \\ a_{n1} & a_{n2} & --- & a_{nn} \end{bmatrix} \cdot \begin{bmatrix} x_1 \\ x_2 \\ \cdot \\ \cdot \\ \cdot \\ x_n \end{bmatrix} = \lambda \begin{bmatrix} x_1 \\ x_2 \\ \cdot \\ \cdot \\ \cdot \\ x_n \end{bmatrix}$$

Thus,

$$a_{11}x_1 + a_{12}x_2 + ---- + a_{1n}x_n = \lambda x_1$$

$$a_{21}x_1 + a_{22}x_2 + ---- + a_{2n}x_n = \lambda x_2$$

$$a_{n1}x_1 + a_{n2}x_2 \quad --- \quad a_{nn}x_n = \lambda x_n$$

There will be n values of λ that will satisfy the above equations. These are the n eigenvalues of the matrix A. The values of X_n that satisfy the above equations for each value of λ_i (i=1, 2, . . . , n) are the eigenvectors associated with λ_i.

From the above set of equations we obtain

$$\lambda X_n - AX_n = 0 \quad \text{or} \quad (\lambda I - A) X_n = 0$$

One obvious solution is the trivial solution $X_n = 0$ ($x_1 = 0, x_2 = 0, \text{---}, x_n = 0$). It will be seen in Chapter 3 that the determinant of the coefficient matrix must be zero for there to be more than one solution. Therefore, for the above equation to have a solution other than the trivial one the following must be satisfied

$$|\lambda I - A| = 0$$

This equation will give an nth degree polynomial that will have n roots. These n roots, or values of λ_i (i=1, 2, ---, n), are the n eigenvalues of the matrix A. Each of these values of λ_i may be used in the equation $AX_n = \lambda X_n$ to obtain values of X. The X vector is an eigenvector of the matrix A.

Example:

$$\text{Let A} = \begin{bmatrix} 3 & 1 \\ 2 & 4 \end{bmatrix}$$

Then $AX = \lambda X$,

$$\begin{bmatrix} 3 & 1 \\ 2 & 4 \end{bmatrix} \begin{bmatrix} x_1 \\ x_2 \end{bmatrix} = \lambda \begin{bmatrix} x_1 \\ x_2 \end{bmatrix}$$

or

$$3x_1 + x_2 = \lambda x_1$$

$$2x_1 + 4x_2 = \lambda x_2$$

and

$$|\lambda I - A| = 0 = \left| \lambda \begin{bmatrix} 1 & 0 \\ 0 & 1 \end{bmatrix} - \begin{bmatrix} 3 & 1 \\ 2 & 4 \end{bmatrix} \right|$$

$$0 = \begin{vmatrix} \lambda-3 & -1 \\ -2 & \lambda-4 \end{vmatrix}$$

$$0 = (\lambda-3)(\lambda-4) - (-1)(-2)$$

$$0 = \lambda^2 - 7\lambda + 12 - 2$$

$$0 = \lambda^2 - 7\lambda + 10$$

$$0 = (\lambda-2)(\lambda-5)$$

The two roots of this polynomial are $\lambda_1 = 2$, $\lambda_2 = 5$, which are the two eigenvalues of A.

For $\lambda_1 = 2$, $AX = \lambda X$ becomes

$$3x_1 + x_2 = 2x_1$$

$$2x_1 + 4x_2 = 2x_2$$

which gives $x_1 = -x_2$. (The reader who has had high school algebra should be able to obtain this solution. A discussion of linear equations is presented in Chapter 3. This discussion will enable the reader to efficiently solve more complicated sets of equations.) If $x_1 = 1$, then $x_2 = -1$ and one eigenvector associated with λ_1 is

$$X = \begin{bmatrix} 1 \\ -1 \end{bmatrix}$$

For $\lambda_2 = 5$

$$3x_1 + x_2 = 5x_1$$

$$2x_1 + 4x_2 = 5x_2$$

which gives $x_1 = \frac{1}{2}x_2$ and one eigenvector associated with λ_2 is

$$X = \begin{bmatrix} 1 \\ 2 \end{bmatrix}$$

Eigenvalues will be used in Chapter 2, Optimization with Calculus, and practical examples will be given in Chapter 2.

HESSIAN MATRIX

The Hessian matrix of a function of multiple variables, $f(X)$, where $X = [x_1, x_2, \cdots, x_n]$, is a matrix of second partial derivatives.

$$H_{f(x)} = \begin{bmatrix} \dfrac{\partial^2 f}{\partial x_1^{\,2}} & \dfrac{\partial^2 f}{\partial x_1\,\partial x_2} & - - - & \dfrac{\partial^2 f}{\partial x_1\,\partial x_n} \\[2ex] \dfrac{\partial^2 f}{\partial x_2\,\partial x_1} & \dfrac{\partial^2 f}{\partial x_2^{\,2}} & - - - & \dfrac{\partial^2 f}{\partial x_2\,\partial x_n} \\[2ex] \cdot & \cdot & - - - & \cdot \\ \cdot & \cdot & - - - & \cdot \\ \cdot & \cdot & - - - & \cdot \\[1ex] \dfrac{\partial^2 f}{\partial x_n\,\partial x_1} & \dfrac{\partial^2 f}{\partial x_n\,\partial x_2} & - - - & \dfrac{\partial^2 f}{\partial x_n^{\,2}} \end{bmatrix}$$

Example:

$$f(X) = 3x_1^{\,2} + 4x_2^{\,2} + x_3^{\,2} - 6x_1 x_2 + 7x_1 x_3 - 2x_2^{\,2} x_3 + 9x_1$$
$$- 5x_2 - 8x_3 + 23$$

$$\frac{\partial f}{\partial x_1} = 6x_1 - 6x_2 + 7x_3 + 9$$

$$\frac{\partial f}{\partial x_2} = 8x_2 - 6x_1 - 4x_2 x_3 - 5$$

$$\frac{\partial f}{\partial x_3} = 2x_3 + 7x_1 - 2x_2{}^2 - 8$$

$$\frac{\partial}{\partial x_1}\left(\frac{\partial f}{\partial x_1}\right) = \frac{\partial^2 f}{\partial x_1{}^2} = +6$$

$$\frac{\partial}{\partial x_2}\left(\frac{\partial f}{\partial x_2}\right) = \frac{\partial^2 f}{\partial x_2{}^2} = 8 - 4x_3$$

$$\frac{\partial}{\partial x_3}\left(\frac{\partial f}{\partial x_3}\right) = \frac{\partial^2 f}{\partial x_3{}^2} = 2$$

$$\frac{\partial^2 f}{\partial x_1\,\partial x_2} = \frac{\partial^2 f}{\partial x_2\,\partial x_1} = -6$$

$$\frac{\partial^2 f}{\partial x_1\,\partial x_3} = \frac{\partial^2 f}{\partial x_3\,\partial x_1} = +7$$

$$\frac{\partial^2 f}{\partial x_2\,\partial x_3} = \frac{\partial^2 f}{\partial x_3\,\partial x_2} = -4x_2$$

$$H_f(X) = \begin{bmatrix} 6 & -6 & 7 \\ -6 & (8-4x_3) & -4x_2 \\ 7 & -4x_2 & 2 \end{bmatrix}$$

The Hessian matrix is useful in optimizing functions of several variables with calculus. This will be illustrated in the next chapter.

SUGGESTED ADDITIONAL READINGS

Campbell, Hugh G. *An Introduction to Matrices, Vectors and Linear Programming.* New York: Appleton-Century-Crofts, 1965.

EXERCISES

For the following matrices:

$$A = \begin{bmatrix} 3 & -1 \\ 2 & 6 \end{bmatrix} \qquad B = \begin{bmatrix} 1 \\ 2 \end{bmatrix} \qquad C = \begin{bmatrix} 1 & 1 \\ 1 & 1 \end{bmatrix}$$

$$D = \begin{bmatrix} 3 & 4 \\ 2 & 0 \\ 1 & 5 \end{bmatrix} \qquad E = \begin{bmatrix} 3 & 1 & 1 \\ 6 & 0 & 0 \\ 9 & 4 & 0 \end{bmatrix} \qquad F = \begin{bmatrix} 6 & 2 & 3 \end{bmatrix}$$

$$G = \begin{bmatrix} 2 & 3 & 9 \\ 1 & 0 & 2 \end{bmatrix}$$

1. Find the following, if possible:

   (a) $A + B$ (g) BI

   (b) $A + C$ (h) $(A+C)B$

   (c) $D + G^T$ (i) FE

   (d) $G^T + D$ (j) $D - G$

   (e) DG (k) $A - C$

   (f) EF (l) GF

2. Calculate:

   (a) $|A|$

   (b) $|C|$

   (c) $|E|$

3. Calculate:

   (a) A^{-1}

   (b) C^{-1}

   (c) E^{-1}

4. Find the eigenvalues of the following, and give one eigenvector associated with each eigenvalue:

   (a) A

   (b) C

5. Give the Hessian matrix of each of the following functions:

   (a) $6x_1 + 5$

   (b) $6x_1 + 4x_2 + 5$

   (c) $3x_1 + 4x_1 x_2 + 6x_2^2 + 7$

(d) $4x_1 + 2x_1 + 3x_1^2 x_2^2 + 7x_1 x_2^2$

(e) $13x_1^2 + 6x_2^2 + 6x_1 + 7x_2 + 8x_1^2 + 13$

(f) $16x_1^2 + 72x_2^4 + 3x_2 + 6x_1^2 x_2 + 9$

(g) $x_1^3 + x_2^2 + x_3^2 + 7x_1 x_2 + 5x_1 x_3 + 14x_2 x_3 + 8x_3$

(h) $x_1^2 + x_2 + x_3 x_2 + 7$

2

OPTIMIZATION
WITH CALCULUS

INTRODUCTION

This chapter presents a basic discussion of optimization with calculus and provides the necessary foundations for nonlinear programming. *The reader who is
interested solely in linear programming may omit this chapter and proceed to
Chapters 3, 4, 5 and 6. Although this material should give additional insight into
linear programming, it is not a prerequisite.* The procedures presented in this
chapter are applicable to linear programming problems; however, the special
procedures of Chapters 3, 4, 5 and 6 are much more efficient in dealing with this
common category of problems. The reader who has no background in calculus,
or who is not interested in optimization with calculus, should read the following
brief section on Maxima, Minima and Saddle Points for general background
information before a perusal of Chapter 7, Search Procedures. No knowledge of
calculus is assumed for Chapter 7.

A mathematical expression, which is referred to as the objective function,
will relate a response to the decision variable(s). As an example, a response may
be inventory costs. The decision variable is order quantity. The equation which
gives inventory costs as a function of order quantity is our objective function. As
another example, our goal may be to maximize the hardness of a steel. If this
response is the function of two decision variables, percentage of alloy 1 and
percentage of alloy 2, the equation which gives hardness as a function of these
alloys is the objective function.

Some values of the decision variables may not be feasible. For example, inventory order quantity may be limited by available funds. The equation which expresses the investment in inventory as a function of order quantity will define this constraint and is referred to as a constraint equation. The percentages of alloy 1 and alloy 2 may be constrained by a limit on the ductility of the steel. An equation that expresses ductility as a function of percentages of alloys 1 and 2 is a constraint equation.

The first step in mathematical programming is to define the decision variables, the objective function and the constraints. The analysis of the objective function, with consideration for the constraints, will reveal the value(s) of the decision variable(s) that gives optimum response. This approach can be applied to a wide variety of fields of human endeavor that require quantitative decisions.

In many cases, engineering, scientific or economic laws and rules are used to formulate the objective function and constraint equations. In other cases, the relationship between the response and the decision variables is unknown, or is so complex that the objective function is intractable. When this is true the search procedures presented in Chapter 7 may be applied.

In this chapter the basic concepts of optimization, maxima, minima, saddle points and convexity are presented initially; then the relatively straightforward problem of optimizing a function of a single variable is discussed. The more tedious problem of optimizing a function of multiple variables is then categorized as to the number and type of constraints, with appropriate solution procedures presented for each category.

MAXIMA, MINIMA

AND

SADDLE POINTS

This discussion of maxima, minima and saddle points is fundamental to understanding mathematical programming. The analyst must be aware of possible characteristics of the objective function and must not confuse saddle points and local optima with the global optimum.

Figure 2-1 shows the function of a single variable, $f(x)$, and illustrates maxima, minima and saddle points. The variable x is the independent variable, or decision variable, while $f(x)$ is the dependent variable, or the response function.

At each of the five points illustrated, a line tangent to the curve is parallel to the abscissa. The first derivative is zero at these points.

Local minima occur at x_2 and x_5. Small perturbations about the local minima of the independent variable result in an increase in $f(x)$.

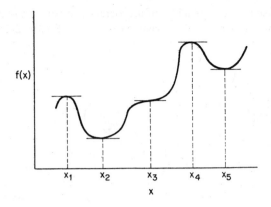

FIGURE 2-1   Function of a Single Variable

$$f(x_2-\Delta) > f(x_2) < f(x_2+\Delta)$$

and

$$f(x_5-\Delta) > f(x_5) < f(x_5+\Delta)$$

for infinitesimally small values of Δ.

Local maxima occur at x_1 and x_4. Small perturbations about these values of the independent variable result in a decrease of $f(x)$.

and
$$f(x_1-\Delta) < f(x_1) > f(x_1+\Delta)$$
$$f(x_4-\Delta) < f(x_4) > f(x_4+\Delta)$$

for infinitesimally small values of Δ.

A saddle point occurs at x_3. A small perturbation in one direction results in an increase in $f(x)$, and in the other direction, a decrease. In addition, a plane drawn tangent to x_3 will have a slope of zero. This is characteristic of saddle points.

The global maximum is the maximum point for the defined range of x and occurs at x_4. There is no value of the independent variable that gives a greater value of $f(x)$ than x_4, for the defined range of the independent variable.

The global minimum occurs at x_2. There is no value of the independent variable that gives a smaller value of $f(x)$ than x_2 for the defined range of the independent variable.

This concept may be easily extended to functions of multiple variables. Functions of two variables are best illustrated by contour maps, analogous to geographical contour maps. As a line of a geographical contour map represents a constant elevation, a line on the illustration below (Figure 2-2) represents a constant value of f(X).

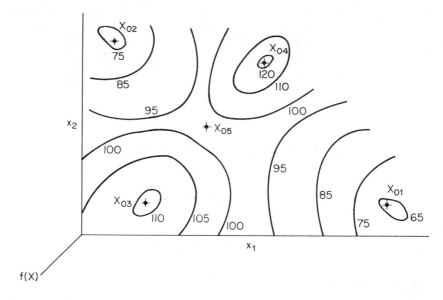

FIGURE 2-2 Function of Two Variables

A capital letter in Figure 2-2 (e.g. X) indicates a point, which consists of two coordinates. Each coordinate is denoted by a lower-case letter, x_1 and x_2, respectively. *Throughout this text, a lower-case letter indicates a single variable, while an upper-case letter indicates a vector, consisting of multiple variables.* For example, f(X) denotes a function expressed in terms of two or more variables.

It can be seen that X_{01} and X_{02} are local minima since perturbations from these points in any direction ($\pm \Delta$ in either or both variables) result in an increase in the response function f(X).

Since all perturbations about X_{03} or X_{04} result in a decrease in the response function, these points are local maxima.

A saddle point occurs at X_{05} since perturbations in some directions give an increase in f(X) and in other directions, a decrease.

The global maximum is at X_{04} because $X_{04} = 120 > X_{03} = 110$ and the global minimum at X_{01} because $X_{01} = 65 < X_{02} = 75$.

At all five points

$$\frac{\partial f}{\partial x_1} = 0 \quad \text{and} \quad \frac{\partial f}{\partial x_2} = 0$$

CONVEXITY

The analyst must determine if his objective functions and constraint equations are concave or convex. In real-world problems, if the objective function or the constraints are not concave or convex, the problem is usually mathematically intractable. A function is *strictly convex* if a line connecting any two points on the function lies completely above the function. The following is a mathematical statement and illustration of this definition for a function of one variable.

$$f[\alpha x_1 + (1-\alpha)x_2] < \alpha f(x_1) + (1-\alpha)f(x_2)$$

where $0 \leqslant \alpha \leqslant 1$.

If the "<" is replaced by "$\leqslant$" the expression will define *convex* functions. The above relationship is illustrated in Fig. 2-3.

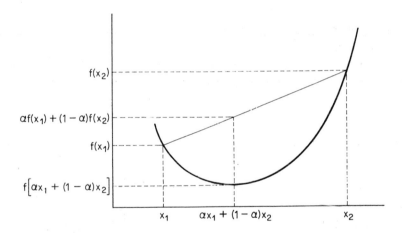

FIGURE 2-3   A Convex Function of One Variable

A function is *strictly convex* if its slope is continually increasing, or $d^2 f/dx^2 > 0$. It is convex if its slope is nondecreasing, or $d^2 f/dx^2 \geqslant 0$.

A function is *strictly concave*, or *strictly convex upward*, if a line connecting any two points on the function lies completely below the function (intersects only at two points). The following is a mathematical statement and illustration of this definition. See Figure 2-4 for graphical illustration.

$$f[\alpha x_1 + (1-\alpha)x_2] > \alpha f(x_1) + (1-\alpha) f(x_2)$$

where $0 \leqslant \alpha \leqslant 1$.

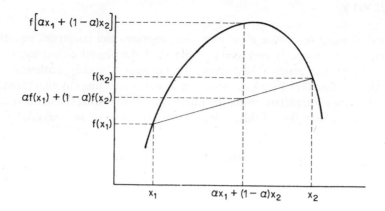

FIGURE 2-4 A Concave Function of One Variable

If the ">" is replaced by the "$\geqslant$", this expression will define *concave* or *convex upward functions.*

A function is *strictly concave* if its slope is continually decreasing, or $d^2f/dx^2 < 0$. It is concave if its slope is nonincreasing or $d^2f/dx^2 \leqslant 0$.

Examples:

(a) $f(x) = x^2 - 6x + 4$

$$\frac{df}{dx} = 2x - 6$$

$$\frac{d^2f}{dx^2} = +2$$

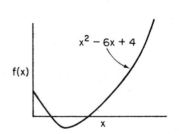

Since $d^2f/dx^2 > 0$, $f(x)$ is strictly convex.

(b) $f(x) = 7x + 4$

$$\frac{df}{dx} = +7$$

$$\frac{d^2f}{dx^2} = 0$$

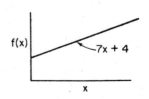

Since $d^2 f/dx^2 = 0$, the function is both convex and concave.

(c) $f(x)$ $= 2x^3 - x^2 + 2x + 5$

$\dfrac{df}{dx}$ $= 6x^2 - 2x + 2$

$\dfrac{d^2 f}{dx^2}$ $= 12x - 2$

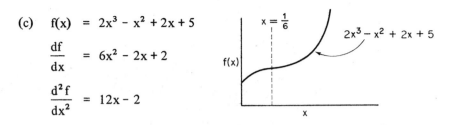

The function $f(x)$ is strictly convex for values of $x > 1/6$ and strictly concave for values of $x < 1/6$. A point of inflection is at $x = 1/6$.

A function of two variables, $f(X)$ where $X = [x_1 \ x_2]$, is strictly convex if

$$f[\alpha X_1 + (1 - \alpha)X_2] < \alpha f(X_1) + (1-\alpha) f(X_2)$$

where X_1 and X_2 are points located by the coordinates given in their respective vectors. A convex function is illustrated in Figure 2-5, and a concave function in Figure 2-6. The function shown earlier in Figure 2-2 is neither concave nor convex.

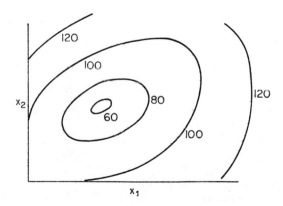

FIGURE 2-5   A Convex Function of Two Variables

To determine convexity or concavity of a function of multiple variables, the eigenvalues of its Hessian matrix should be examined:

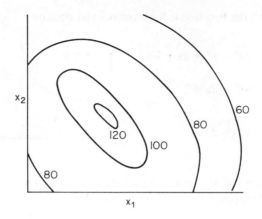

FIGURE 2-6   A Concave Function of Two Variables

(a)   If all eigenvalues of the Hessian are negative the function is strictly concave.

(b)   If all eigenvalues of the Hessian are positive the function is strictly convex.

(c)   If some eigenvalues are positive and some negative, or if some are zero, the function is neither strictly concave nor strictly convex.

Example:

$$x_1{}^2 + x_2{}^2 - 6x_1 + 3x_2 - 7$$

$$\frac{\partial f}{\partial x_1} = 2x_1 - 6 \qquad\qquad \frac{\partial f}{\partial x_2} = 2x_2 + 3$$

$$\frac{\partial^2 f}{\partial x_1{}^2} = +2 \qquad\qquad \frac{\partial^2 f}{\partial x_2{}^2} = +2$$

$$\frac{\partial^2 f}{\partial x_2\, \partial x_1} = \frac{\partial^2 f}{\partial x_1\, \partial x_2} = 0$$

$$H = \begin{bmatrix} 2 & 0 \\ 0 & 2 \end{bmatrix}$$

$$|\lambda I - H| = \begin{vmatrix} \lambda-2 & 0 \\ 0 & \lambda-2 \end{vmatrix} = 0$$

$$(\lambda - 2)(\lambda - 2) = 0 \quad \text{or} \quad \lambda_1 = +2, \lambda_2 = +2$$

Since both eigenvalues are positive, the function $f(X)$ is strictly convex.

The following are brief statements of some properties of concave and convex functions.

1. A local minimum of a convex function is also the global minimum, and a local maximum of a concave function is also a global maximum.

2. A straight line is both concave and convex.

3. The sum of (strictly) convex functions is (strictly) convex, and the sum of (strictly) concave functions is (strictly) concave.

4. If $f(X)$ is a convex function and k is a constant, then
   (a) $kf(X)$ is convex if $k > 0$.
   (b) $kf(X)$ is concave if $k < 0$.

OPTIMIZATION OF

A FUNCTION OF

ONE VARIABLE

Care should be taken when optimizing a mathematical function. The analyst should attempt to develop intuitive judgment as to the behavior of the function. This may be accomplished with a function of one variable by supplementing mathematical calculation with graphical analysis. In the following discussion, stationary points will be determined, second derivatives will be examined to check for convexity or concavity, and tests will be recommended to evaluate stationary points as local maxima, local minima or saddle points. This information will assist in graphing the function.

necessary conditions
for a stationary point

A stationary point is a value of the independent variable at which the slope of the function is zero. Any stationary point, x_0, must have the property that

$$\left. \frac{df}{dx} \right|_{x_0} = 0$$

To determine the stationary points, the equation $df/dx = 0$ must be solved.

**sufficient conditions
for a stationary
point**

To determine convexity or concavity, the second derivatives can be examined. If $d^2f/dx^2 > 0$ for all values of x, f(x) is convex and the stationary point is a global minimum. If $d^2f/dx^2 < 0$ for all values of x, f(x) is concave and the stationary point is a global maximum.

    If the function is neither concave nor convex, the following test may be used to classify stationary points.

1. Find the first nonzero higher-order derivative. This will be the nth derivative of f(X).

$$\frac{d^nf}{dx^n}\bigg|_{x_0} \neq 0$$

where x_0 is the stationary point.

2. If n is odd, x_0 is a saddle point.

3. If n is even, x_0 is a local maximum or a local minimum.

(a)   If $\dfrac{d^nf}{dx^n}\bigg|_{x_0} < 0$

         x_0 is a local maximum.

(b)   If $\dfrac{d^nf}{dx^n}\bigg|_{x_0} > 0$

         x_0 is a local minimum.

Example:

In a chemical reaction the percent of an undesirable by-product is found to be a function of time. By using mathematical curve-fitting techniques with several data points, the following equation is developed as an approximation of this relationship. The following is a mathmatical analysis of this function, where the goal is to minimize f(t).

$$\text{Min } f(t) = 3t^4 - 4t^3 + 2$$

where

t   = hours

f(t)   = percent of by-product

$$\frac{df}{dt} = 12t^3 - 12t^2 = 12t^2(t - 1) = 0$$

Solving this equation, we find that $t_{01} = 0$ and $t_{02} = 1$ are the two stationary points. Examining the second derivative,

$$\frac{d^2f}{dt^2} = 36t^2 - 24t$$

it can be shown that d^2f/dt^2 is positive for values of t from $-\infty$ to 0, and from $+2/3$ to ∞, and negative for values of t from 0 to $+2/3$. Therefore, the entire function is neither concave nor convex.

Examining the stationary point $t_{01} = 0$:

$$\frac{d^2f}{dt^2}\bigg|_{t_{01}} = 36t^2 - 24t = 0$$

$$\frac{d^3f}{dt^3}\bigg|_{t_{01}} = 72t - 24 = -24$$

Since the first nonzero higher-order derivative occurs at an odd value of n, n=3, there is a saddle point at $t_{01} = 0$.

Examining the stationary point at $t_{02} = 1$:

$$\frac{d^2f}{dt^2}\bigg|_{t_{02}} = 36t^2 - 24t = +12$$

Since n is even, n = 2 and $d^2f/dt^2 = 12 > 0$, a local minimum occurs at t_{02} = 1.

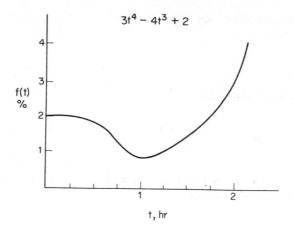

FIGURE 2-7

Figure 2-7 is a graphical illustration of this function. From the plot of $f(t)$ in Figure 2-7, we see that t=1 is also the global minimum. Therefore, to minimize the percent of our undesirable component, we should operate at a reaction time of one hour.

OPTIMIZATION
OF A FUNCTION
OF MULTIPLE
VARIABLES

Functions of multiple variables are more difficult to analyze than are functions of a single variable since graphical illustration is difficult or impossible, and more calculations are involved in mathematical analysis. If a convex function is to be minimized, the stationary point is the global minimum and analysis is relatively straightforward. A similar situation exists in maximizing a concave function. For other cases, a location of the global optimum may not be possible by using calculus. Some problems can be optimized by inspection, such as Max $f(X) = x_1^2 + x_2^2 + 3x_1x_2 + 6x_1 - 5$, which is obviously maximized at $X = [+\infty, +\infty]$. For other functions, application of search procedures may be required to develop an estimate of the optimum. This topic is discussed in Chapter 7.

**unconstrained
optimization**

A necessary condition for a stationary point of the function f(X) is that each partial derivative equal zero. In other words, each element of the gradient vector must equal zero where the gradient vector of f(X), $\Delta_X f$, is defined as follows:

$$\Delta_X f = \begin{bmatrix} \dfrac{\partial f}{\partial x_1} \\[2mm] \dfrac{\partial f}{\partial x_2} \\[1mm] \cdot \\ \cdot \\ \cdot \\[1mm] \dfrac{\partial f}{\partial x_n} \end{bmatrix}$$

This is illustrated graphically in Figure 2-8.

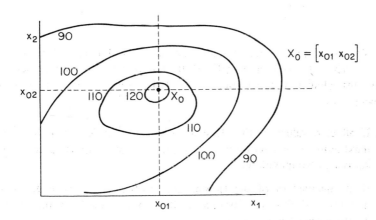

FIGURE 2-8 The Stationary Point of a Concave Function

If a cross section of f(X) is taken at the x_{01} line, then the highest point on this line is at X_0, and the slope of the cross section at this point is zero, $\partial f/\partial x_2 = 0$. Examining the x_{02} line it can be seen similarly that $\partial f/\partial x_1 = 0$ at X_0, and that X_0 satisfies the necessary conditions for a stationary point.

The response surface illustrated in Figure 2-9 has a point, X_1, that satisfies the *necessary* conditions for a stationary point, but yet it is not a stationary point. The planes defined by x_{11} and x_{12} result in a strictly concave contour so that $\Delta_x f = 0$ at X_1. This point is not stationary, however, because a plane tangent to the response surface at X_1 has a positive slope. In Figure 2-9, X_0 is a true stationary point that is also the global maximum.

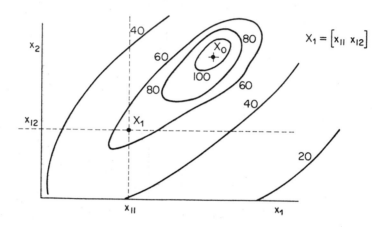

FIGURE 2-9 A Point Satisfying the Necessary Conditions for a Stationary Point

To check the sufficient conditions at a stationary point, X_0, of the function f(X), the Hessian matrix of f(X) may be evaluated at X_0 and its eigenvalues calculated. The stationary point then may be classified by using the following rules:

1. If all eigenvalues of the Hessian are negative at X_0, then X_0 is a local maximum. If all eigenvalues are negative for all possible values of X, then X_0 is a global maximum.

2. If all eigenvalues of the Hessian are positive at X_0, then X_0 is a local minimum. If all eigenvalues are positive for all possible values of X, then X_0 is a global minimum.

3. If some eigenvalues of the Hessian at X_0 are positive and some negative, or if some are zero, the stationary point, X_0, is neither a local maximum nor a local minimum.

Example:

The yield of a chemical reaction is the actual production as a percent of that which is theoretically possible. In a large commercial operation, production is found to be a function of two catalysts x_1 and x_2, where it is our objective to maximize yield (%): $f(X) = 60 + 8x_1 + 2x_2 - x_1^2 - \frac{1}{2}x_2^2$

$$\frac{\partial f}{\partial x_1} = 8 - 2x_1 = 0$$

$$\frac{\partial f}{\partial x_2} = 2 - x_2 = 0$$

Solving these equations, we get a stationary point of $X = [4, 2]$. Now we must determine the nature of this stationary point.

$$\frac{\partial^2 f}{\partial x_1^2} = -2 \qquad \frac{\partial^2 f}{\partial x_2^2} = -1 \qquad \frac{\partial^2 f}{\partial x_1 \partial x_2} = 0$$

$$H = \begin{bmatrix} -2 & 0 \\ 0 & -1 \end{bmatrix}$$

$$|\lambda I - H| = \begin{vmatrix} \lambda+2 & 0 \\ 0 & \lambda+1 \end{vmatrix} = (\lambda+2)(\lambda+1) = 0$$

We see that the values of λ do not depend on X and that $\lambda_1 = -2, \lambda_2 = -1$. Since the eigenvalues are both negative, $f(X)$ is concave and $x_1 = 4\%$, $x_2 = 2\%$ will give the global maximum yield of $f(X) = 78.0\%$.

The following example is included to illustrate location and classification of stationary points for a nonconvex function.

Example:

Locate the stationary points of $f(X)$ and determine whether they are local maxima, local minima, or neither.

$$f(X) = x_1{}^3 - x_1 x_2 + x_2{}^2 - 2x_1 + 3x_2 - 4$$

1. $$\frac{\partial f}{\partial x_1} = 3x_1{}^2 - x_2 - 2 = 0$$

2. $$\frac{\partial f}{\partial x_2} = -x_1 + 2x_2 + 3 = 0$$

3. From 2,

$$x_1 = 2x_2 + 3$$

4. From 1 and 3

$$3(2x_2 + 3)^2 - x_2 - 2 = 0$$

$$3(4x_2{}^2 + 12x_2 + 9) - x_2 - 2 = 0$$

$$12x_2{}^2 + 35x_2 + 25 = 0$$

$$(3x_2 + 5)(4x_2 + 5) = 0$$

$$x_2 = -5/3 \quad \text{or} \quad x_2 = -5/4$$

From 3

$$\text{if } x_2 = -5/3, \; x_1 = -1/3$$

$$\text{if } x_2 = -5/4, \; x_1 = +1/2$$

and the two stationary points are

$$X_{01} = [-1/3, -5/3]$$

and

$$X_{02} = [+1/2, -5/4]$$

The Hessian of f(X) is

$$\frac{\partial^2 f}{\partial x_1{}^2} = 6x_1 \qquad \frac{\partial^2 f}{\partial x_2{}^2} = +2$$

$$\frac{\partial f}{\partial x_1 \partial x_2} = \frac{\partial f}{\partial x_2 \partial x_1} = -1$$

$$H_{f(x)} = \begin{bmatrix} 6x_1 & -1 \\ -1 & +2 \end{bmatrix}$$

Now to calculate the eigenvalues of the Hessian at each of the stationary points:

$$[\lambda I - H] = \begin{bmatrix} \lambda - 6x_1 & +1 \\ +1 & \lambda - 2 \end{bmatrix}$$

At $X_{01} = [-1/3, -5/3]$

$$|\lambda I - H| = \begin{vmatrix} \lambda + 2 & +1 \\ +1 & \lambda - 2 \end{vmatrix} = (\lambda + 2)(\lambda - 2) - 1 = 0$$

$$\lambda^2 - 5 = 0$$

$$\lambda^2 = 5$$

$$\lambda = +\sqrt{5} \quad \text{and} \quad \lambda = -\sqrt{5}$$

Since one eigenvalue is positive and the other negative, X_{01} is neither a local maximum nor a local minimum.

At $X_{02} = [+1/2, -5/4]$

$$|\lambda I - H| = \begin{vmatrix} \lambda - 3 & +1 \\ +1 & \lambda - 2 \end{vmatrix} = (\lambda - 3)(\lambda - 2) - 1 = 0$$

$$\lambda^2 - 5\lambda + 5 = 0$$

$$\frac{-b \pm \sqrt{b^2 - 4ac}}{2a} = \frac{+5 \pm \sqrt{25 - 20}}{2}$$

the solution to this equation is

$$\lambda_1 = \frac{5 + \sqrt{5}}{2} \qquad \lambda_2 = \frac{5 - \sqrt{5}}{2}$$

Since both of these eigenvalues are positive, X_{02} is a local minimum. Notice that when

$$X = [0, +\infty] , f(x) = +\infty$$

and when

$$X = [-\infty, 0] , f(x) = -\infty$$

so that the global maximum and global minimum responses occur at extreme values of the decision variables.

**optimization of a
function of multiple
variables subject to
an equality
constraint**

A function of multiple variables, $f(X)$, is to be optimized subject to a constraining function of multiple variables, $g(X)$. This equality constraint may or may not be linear. This may be stated mathematically as

Max (or Min) $f(X)$

Subject to $g(X) = 0$

In two variables as shown in Fig. 2-10, $f(X)$ is analogous to the contour mapping of a hill and $g(X)$ to a fence on the hill. The objective is to locate the point on the fence that lies at the highest, or lowest, elevation. Note that the objective function, $f(X)$, is not necessarily linear and the constraint equation, $g(X)$, also may not be linear. The rest of this chapter provides a powerful technique for dealing with nonlinear optimization problems.

One means of dealing with this problem is through Lagrangian multipliers. First, the Lagrangian function is formulated as follows.

$$h\,(X, \lambda) = f(X) - \lambda\,g(X)$$

When $g(X) = 0$, the optimization of $h(X,\lambda)$ gives the same results as the optimization of $f(X)$.

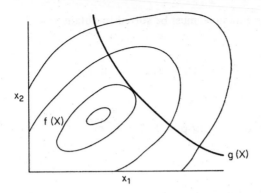

FIGURE 2-10 A Function of Two Variables Subject to a Single Equality Constraint

The conditions necessary for locating a stationary point are:

$$\frac{\partial h\,(X,\lambda)}{\partial x_i} = 0 \qquad \text{for } i = 1, 2, \ldots, n$$

$$\frac{\partial h\,(X,\lambda)}{\partial \lambda} = 0$$

If more than one equality constraint exists, the problem may be stated as

$$\text{Max (or Min)} \quad f(X)$$

$$
\begin{aligned}
\text{Subject to} \quad g_1(X) &= 0 \\
g_2(X) &= 0 \\
&\cdot \\
&\cdot \\
&\cdot \\
g_m(X) &= 0
\end{aligned}
$$

Then,

$$h(X,\lambda) = f(X) - \lambda_1 g_1(X) - \lambda_2 g_2(X) - \ldots - \lambda_m g_m(X)$$

and the following equations must be solved to obtain a solution.

$$\frac{\partial h(X,\lambda)}{\partial x_i} = 0 \qquad \text{for } i = 1, 2, \ldots, n$$

$$\frac{\partial h(X,\lambda)}{\partial \lambda_p} = 0 \qquad \text{for } p = 1, 2, \ldots, m$$

$$K_{ij} = \left. \frac{\partial^2 h(X,\lambda)}{\partial x_i \partial x_j} \right|_{X_0} \qquad \text{for } i, j = 1, 2, \ldots, n$$

$$L_{pi} = \frac{\partial g_p(X)}{\partial x_i} \qquad \text{where } p = 1, 2, \ldots, m \text{ and } i = 1, 2, \ldots, n$$

The solution of linear systems of equations is treated in Chapter 4.

To check for sufficiency, at X_0, the following test may be applied. (The proof of this test will not be given here, since it is too advanced for this treatment. A practical application, however, is within the scope of Chapter 2.) Evaluation of the determinant, Δe, as shown below, will give a polynomial in e of (n–m) order where there are n variables and m constraining equations. Let

$$\Delta e = \begin{vmatrix}
K_{11}-e & K_{12} & \cdots\cdots & K_{1n} & L_{11} & \cdots & L_{m1} \\
K_{21} & K_{22}-e & \cdots\cdots & K_{2n} & L_{12} & \cdots & L_{22} \\
\cdot & \cdot & \cdots\cdots & \cdot & \cdot & \cdots & \cdot \\
\cdot & \cdot & \cdots\cdots & \cdot & \cdot & \cdots & \cdot \\
\cdot & \cdot & \cdots\cdots & \cdot & \cdot & \cdots & \cdot \\
K_{n1} & K_{n2} & \cdots\cdots & K_{nn}-e & L_{1n} & \cdots & L_{mn} \\
L_{11} & L_{12} & \cdots\cdots & L_{1n} & 0 & \cdots & 0 \\
\cdot & \cdot & \cdot & \cdot & \cdot & \cdot & \cdot \\
\cdot & \cdot & \cdot & \cdot & \cdot & \cdot & \cdot \\
\cdot & \cdot & \cdot & \cdot & \cdot & \cdot & \cdot \\
L_{m1} & L_{m2} & \cdots\cdots & L_{mn} & 0 & \cdots & 0
\end{vmatrix}$$

If each root of e in this equation is negative, the point X_0 is a local maximum. If all roots are positive, X_0 is a local minimum. If some are positive and some negative, X_0 is neither a local maximum nor a local minimum.

It should be noted that this analysis does not require convexity of the objective function, or any constraining function.

Example:

An advertising firm has developed a coordinated program for two of our products, x_1 and x_2. Based on experience with these products, they estimate an increased profit of

$$f(X) = -\tfrac{1}{2}x_1{}^2 - \tfrac{1}{2}x_2{}^2 + x_1 x_2 + 3x_2$$

where x_i is the advertising expenditure on product i (i = 1,2). (Suppose that $f(X)$ and x_i are in units of hundreds of thousands of dollars.)

The company has decided to spend exactly \$300,000 on advertising. We now want to allocate this money between the two products and estimate increased profits.

$$\text{Max } f(X) = -\tfrac{1}{2}x_1{}^2 - \tfrac{1}{2}x_2{}^2 + x_1 x_2 + 3x_2$$

$$\text{Subject to } x_1 + x_2 = 3 \qquad \text{or} \qquad x_1 + x_2 - 3 = 0$$

First we formulate the Lagrangian function and locate the stationary points.

$$h(X,\lambda) = -\tfrac{1}{2}x_1{}^2 - \tfrac{1}{2}x_2{}^2 + x_1 x_2 + 3x_2 - \lambda (x_1 + x_2 - 3)$$

$$\frac{\partial h}{\partial x_1} = -x_1 + x_2 - \lambda = 0$$

$$\frac{\partial h}{\partial x_2} = -x_2 + x_1 + 3 - \lambda = 0$$

$$\frac{\partial h}{\partial \lambda} = -x_1 - x_2 + 3 = 0$$

Solving these equations we get $\lambda = 1.5$, $x_1 = 0.75$, $x_2 = 2.25$. Now we determine if this is a maximum.

$$\frac{\partial^2 h}{\partial x_1{}^2} = -1 \qquad \frac{\partial^2 h}{\partial x_2{}^2} = -1 \qquad \frac{\partial^2 h}{\partial x_1 \partial x_2} = +1 \qquad \frac{\partial g}{\partial x_1} = +1 \qquad \frac{\partial g}{\partial x_2} = +1$$

$$\Delta e = \begin{vmatrix} -1-e & +1 & +1 \\ +1 & -1-e & +1 \\ +1 & +1 & 0 \end{vmatrix}$$

Expanding along the last column, we have

$$\Delta e = \begin{vmatrix} +1 & -1-e \\ +1 & +1 \end{vmatrix} - \begin{vmatrix} -1-e & +1 \\ +1 & +1 \end{vmatrix}$$

$$\Delta e = 1 + 1 + e - (-1 - e - 1)$$

$$= 4 + 2e = 0 \qquad \text{or} \qquad e = -2$$

Since the only root of e is negative, our stationary point is a local maximum. Since e does not depend on values of X, this local maximum is also the global maximum and our solution is:

Spend $75,000 on advertising product 1 and $225,000 for product 2

The increased profit will be

$$f(X) = -1/2(0.75)^2 - 1/2(2.25)^2 + 0.75(2.25) - 3(2.25)$$

$$= 5.625, \text{ or } \$562,500$$

**optimization of a
function of multiple
variables subject to a
single inequality
constraint**

This type of problem is similar to that of the previous section with the exception that the constraint is stated as an inequality.

$$\text{Max (or Min) } f(X)$$

$$\text{Subject to } g(X) \leqslant 0, \text{ or } g(X) \geqslant 0$$

In two variables as shown in Figure 2-11, $f(X)$ is analogous to the contour mapping of a hill and $g(X)$ to a fence on the hill. The objective is to locate the point on the fence, or in the field enclosed by the fence, which lies at the highest, or lowest, elevation. This set of acceptable solutions is referred to as the feasible region. In these illustrations, it is assumed that $f(X)$ is to be maximized. The unconstrained optimum of $f(X)$ is shown as X_1, the most desirable point on $g(X)$ as X_2, and the constrained optimum as X_3.

The concept of convex functions was discussed earlier in this chapter. We would like to introduce the concept of a convex region here, in connection with inequality constraints. A region is convex if a straight line connecting any two points in the region lies entirely in the region. In Figure 2-11, the feasible region is convex in Figure 2-11a and 2-11c, and is not convex in Figure 2-11b and 2-11d. If all constraints are straight lines, the feasible region will automatically be convex. The procedures of this section and the next are applicable only if the feasible region is convex. If the feasible region is not convex, the search procedures discussed in Chapter 7 should be applied.

If $f(X)$ is convex for minimizing (concave for maximizing) and the problem is stated in either of the following formats,

Max $f(X)$	Min $f(X)$
Subject to $g(X) \leqslant 0$	Subject to $g(X) \geqslant 0$
($g(X)$ is convex)	($g(X)$ is concave)

the following conclusions may be reached about the solution to the Lagrangian function $h(X,\lambda) = f(X) - \lambda g(X)$—the preceding conditions insure a convex feasible region:

1. If $\lambda \geqslant 0$, the solution obtained is also the solution subject to the inequality constraint.

2. If $\lambda < 0$, determine the maximum or minimum without regard to the constraint. This will be the solution to the problem.

In most analyses of these problems, it is assumed that $f(X)$ is convex (or concave, if maximizing) when dealing with inequality constraints. If $f(X)$ is not convex there is no general procedure that will guarantee location of the global optimum. To gather information in evaluating the nonconvex problem, stationary points of $f(X)$ may be located, the response evaluated at some extreme

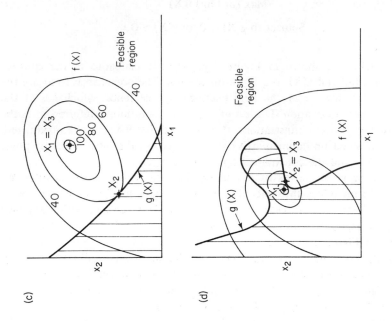

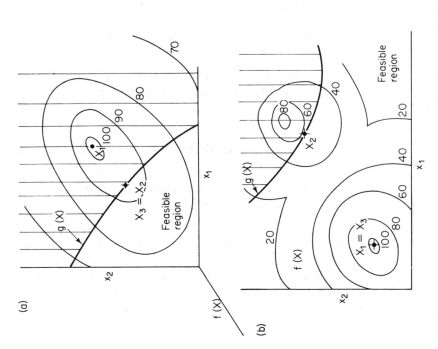

FIGURE 2-11 Function of Multiple Variables Subject to a Single Inequality Constraint

values of the independent variables, and the most desirable point satisfying the constraint then located with a Lagrangian multiplier approach. When this information is compiled, a solution may be obvious if the problem is relatively simple. For more complex problems, an estimate of the optimum may be made, or search procedures (Chapter 7) may be applied.

Example 1:

In the example used with unconstrained functions of multiple variables, the following equation for the yield of a chemical plant is given

$$f(X) = 60 + 8 x_1 + 2x_2 - x_1^2 - \tfrac{1}{2}x_2^2$$

where x_1 and x_2 are the percentages of two catalysts. We found the function to be convex and the unconstrained optimum at $X = [4,2]$. If we impose a constraint on this problem that pressure must be less than or equal to 160 pounds per square inch (psi), where the equation for pressure is given by

$$P(X) = 40x_1 + 20\,x_2 + 20 \leqslant 160$$

then we may obtain a solution by using a Lagrangian multiplier approach.

$$g(X) = 40x_1 + 20x_2 + -140 \leqslant 0$$

$$h(X, \lambda) = 60 + 8x_1 + 2x_2 - x_1^2 - \tfrac{1}{2}x_2^2 - \lambda (40x_1 + 20x_2 - 140)$$

$$\frac{\partial h}{\partial x_1} = 8 - 2x_1 - 40\lambda = 0$$

$$\frac{\partial h}{\partial x_2} = 2 - x_2 - 20\lambda = 0$$

$$\frac{\partial h}{\partial \lambda} = -40x_1 - 20x_2 + 140 = 0$$

Solving these equations, we get $X = [3,1]$, $\lambda = 0.05$. Since λ is positive, $f(X)$ is convex and $g(X)$ is linear, this is the optimum of our constrained problem. Therefore, we would recommend use of 3% of catalyst x_1 and 1% of catalyst x_2.

Example 2:

$$\text{Maximize } f(X) = 5x_1 - 2x_1{}^2 + 3x_1x_2 - 2x_2{}^2$$

$$\text{Subject to } x_1 + x_2 \leqslant 2$$

First let us determine if the objective function is concave:

$$\frac{\partial f}{\partial x_1} = 5 - 4x_1 + 3x_2$$

$$\frac{\partial f}{\partial x_2} = 3x_1 - 4x_2$$

$$\frac{\partial^2 f}{\partial x_1{}^2} = -4$$

$$\frac{\partial^2 f}{\partial x_2{}^2} = -4$$

$$\frac{\partial^2 f}{\partial x_1 \partial x_2} = +3$$

$$H_{f(X)} = \begin{bmatrix} -4 & +3 \\ +3 & -4 \end{bmatrix}$$

$$|\lambda I - H_{f(X)}| = \begin{vmatrix} \lambda+4 & -3 \\ -3 & \lambda+4 \end{vmatrix}$$

$$= (\lambda+4)^2 - 9 = \lambda^2 + 8\lambda + 7 = 0$$

$$\lambda_1 = -7$$

$$\lambda_2 = -1$$

Since both eigenvalues of the Hessian are negative, the objective function is concave. Thus, procedures presented in this chapter are applicable.

The Lagrangian is

$$h(X,\lambda) = 5x_1 - 2x_1{}^2 + 3x_1x_2 - 2x_2{}^2 - \lambda(x_1 + x_2 - 2)$$

$$\frac{\partial h}{\partial x_1} = 5 - 4x_1 + 3x_2 - \lambda = 0 \qquad \text{(Eq. 1)}$$

$$\frac{\partial h}{\partial x_2} = 3x_1 - 4x_2 - \lambda = 0 \qquad \text{(Eq. 2)}$$

$$\frac{\partial h}{\partial \lambda} = -x_1 - x_2 + 2 = 0 \qquad\qquad \text{(Eq. 3)}$$

Solving these equations:

$$x_1 + x_2 - 2 = 0 \quad \text{ or } \quad x_1 = 2 - x_2$$

Upon substitution into (Eq. 2)

$$3(2-x_2)-4(x_2) - \lambda = 0$$
$$\lambda = 6 - 7x_2$$
$$x_2 = 6/7 - (1/7)\lambda$$

From (Eq. 3),

$$x_1 = 2 - x_2$$

so that

$$x_1 = 2 - (6/7 - \lambda/7)$$
$$x_1 = 8/7 + \lambda/7$$

Upon substitution of these expressions for x_1 and x_2 into (Eq. 1), we can solve for a value of λ:

$$5 - 4(8/7 + \lambda/7) + 3(6/7 - \lambda/7) - \lambda = 0$$
$$3 - 2\lambda = 0$$
$$\lambda = 3/2$$

Finally,

$$x_2 = 6/7 - 1/7\lambda = 6/7 - 1/7(3/2) = \ 9/14$$
$$x_1 = 8/7 + 1/7\lambda = 8/7 + 1/7(3/2) = 19/14$$

The values of x_1, x_2 and λ are as follows.

$$x_1 = 19/14 \qquad f(X^*) = 4.89$$

$$x_2 = 9/14$$

$$\lambda = 3/2$$

Since λ is positive, the constraint is binding.

The graphical approach can be used for quick verification as shown in Figure 2-12.

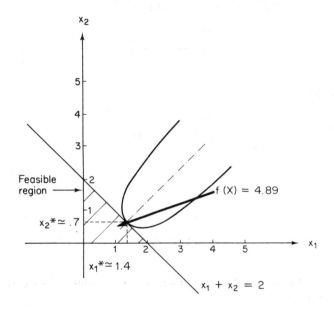

FIGURE 2-12 Graphical Illustration of a Function of Two Variables Having One Inequality Constraint

optimization of a function of multiple variables subject to multiple inequality constraints

This problem is similar to that of the previous section with the exception that there is more than one constraint.

$$\text{Max (or Min) } f(X)$$

$$\text{Subject to } g_1\,(X) \;\leqslant\; 0 \text{ (or } g_i(X) \geqslant 0)$$

$$g_2\,(X) \;\leqslant\; 0$$

$$\cdot$$
$$\cdot$$
$$\cdot$$

$$g_m(X) \;\leqslant\; 0$$

This is analogous to several fences on a hill. The objective is to locate the highest point on the hill without crossing any fence. A concave objective function is illustrated in Figure 2-13.

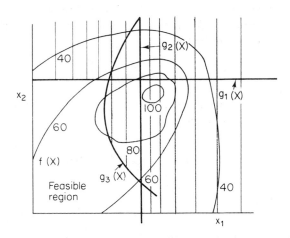

FIGURE 2-13 A Function of Multiple Variables Subject to Multiple In-equality Constraints

If $f(X)$ is convex for minimizing (concave for maximizing), the problem is stated in either of the following formats:

Max $f(X)$	Min $f(X)$
Subject to $g_1(X) \leqslant 0$	Subject to $g_1(X) \geqslant 0$
$g_2(X) \leqslant 0$	$g_2(X) \geqslant 0$
$\cdot$	$\cdot$
$\cdot$	$\cdot$
$\cdot$	$\cdot$

$$g_p(X) \leqslant 0 \qquad\qquad g_p(X) \geqslant 0$$

(all g(X) are convex)   (all g(X) are concave)

A Lagrangian multiplier approach may be used to investigate this function: $h(X,\lambda) = f(X) - \lambda_1 g_1(X) - \lambda_2 g_2(X) \cdot \cdot \cdot \cdot -\lambda_p g_p(X)$. (The above conditions insure a convex feasible region.)

If any $\lambda_i < 0$, the constraint associated with that λ_i is not active and another solution should be obtained disregarding inactive constraints.

If f(X) is not convex, this may become an extremely tedious problem. The analyst may not know how many constraints are active, or truly binding, and how many are inactive. To gather insight into the problem, the analyst may evaluate the problem m times with f(X) subject to each constraint separately, locate stationary points of f(X), and evaluate f(X) at different values of extremes in the independent variables. All of this may only prepare the analyst to make an estimate of the global optimum.

Example:

Let us assume that the example problem of the previous section is also subject to a constraint on temperature, where

$$g(X) = 50x_1 + 35x_2 + 40 \leqslant 240$$

Our problem becomes

$$\text{Maximize } f(X) \quad = \quad 60 + 8x_1 + 2x_2 - x_1{}^2 - \tfrac{1}{2}x_2{}^2$$

$$\text{Subject to} \qquad 40x_1 + 20x_2 - 140 \leqslant 0$$

$$50x_1 + 35x_2 - 200 \leqslant 0$$

$$h(X,\lambda) \quad = \quad 60 + 8x_1 + 2x_2 - x_1{}^2 - \tfrac{1}{2}x_2{}^2$$

$$-\lambda_1 (40x_1 + 20x_2 - 140)$$

$$-\lambda_2 (50x_1 + 35x_2 - 200)$$

$$\frac{\partial h}{\partial x_1} = 8 - 2x_1 - 40\lambda_1 - 50\lambda_2 = 0$$

$$\frac{\partial h}{\partial x_2} = 2 - x_2 - 20\lambda_1 - 35\lambda_2 = 0$$

$$\frac{\partial h}{\partial \lambda_1} = -40x_1 - 20x_2 + 140 = 0$$

$$\frac{\partial h}{\partial \lambda_2} = -50x_1 - 35x_2 + 200 = 0$$

Solving these equations we find that $x_1 = 2.25$, $x_2 = 2.5$, $\lambda_1 = 0.369$, $\lambda_2 = -0.225$

Since λ_2 is negative, the second constraint is not active. The constraint on temperature should be deleted and the problem solved subject only to the constraint on pressure. This was done in the previous section, where the solution $x_1 = 3$, $x_2 = 1$, $\lambda_1 = 0.05$ is also the solution to this problem.

Kuhn-Tucker conditions

The Kuhn-Tucker Conditions are the necessary conditions for a point to be a local optimum of a function subject to inequality constraints. This is a precise mathematical statement of the procedures used in the previous two sections. If we wish to maximize $f(X)$ subject to $g_1(X) \leq 0, g_2(X) \leq 0, \cdots, g_p(X) \leq 0$, where $X = [x_1 \ x_2 \cdots x_n]$, then the Kuhn-Tucker Conditions for $X^* = [x_1^* x_2^* \cdots x_n^*]$ to be a local maximum are

$$\frac{\partial f(X)}{\partial x_i} - \sum_{j=1}^{P} \lambda_j \frac{\partial g_j(X)}{\partial x_i} = 0 \qquad \text{for } i = 1, 2, \text{---}, n \text{ at } X = X^*$$

and

$$\lambda_j g_j(X) = 0 \qquad g_j(X) \leq 0$$
$$\lambda_j \geq 0 \qquad \text{for } j = 1, 2, \ldots, p \text{ at } X = X^*$$

These are sufficient conditions for a global maximum if $f(X)$ is concave and the constraints form a convex set.

From the example in the section, "Optimization of a Function of Multiple Variables Subject to a Single Inequality Constraint," the solution of

$$\text{Maximize } f(X) = 60 + 8x_1 + 2x_2 - x_1^2 - 1/2 x_2^2$$

$$\text{Subject to } g(X) = 40x_1 + 20x_2 + 20 \leqslant 160$$

is $X = [3, 1]$, $\lambda = 0.05$.

The reader can verify that this satisfies the Kuhn-Tucker Conditions and is also sufficient since $f(X)$ is concave and the $g(X)$s form a convex set.

SUGGESTED ADDITIONAL READINGS

Beveridge, Gordon S.G., and Schechter, Robert S. *Optimization: Theory and Practice*. New York: McGraw Hill, 1970.

Cooper, Leon, and Steinberg, David. *Introduction to Methods of Optimization*. Philadelphia: W. B. Saunders Co., 1970.

Gue, Ronald L., and Thomas, Michael E. *Mathematical Methods in Operations Research*. New York: Macmillan Company, 1968.

EXERCISES

Note: Variables may assume negative as well as positive values.

1. Locate the stationary points, if any, and the global maximum and global minimum of the following functions of one variable. Determine if each function is concave or convex.

    (a) $f(x) = 6x^4 + 8x^3$

    (b) $f(x) = 3x^2 - 5x + 6$

    (c) $f(x) = -80/x + 10x^2$

    (d) $f(x) = x^2/(1 + x^2)$

    (e) $f(x) = x^3 + 6x$

    (f) $f(x) = x^3 - x - x^2$

2. Locate the desired optimum of the following functions of multiple variables:

    (a) Minimize $f(X) = 2x_1^2 + x_2^2 + x_1x_2 - 6x_1$

    (b) Minimize $f(X) = x_1^2 + x_2^2 + x_1x_2 + x_2^3$

    (c) Minimize $f(X) = x_1^2 + x_2^2 - 4x_1 - x_2$

(d) Maximize $f(X) = x_1^2 + x_2^2 - 4x_1 - x_2$

(e) Maximize $f(X) = 8x_1^3 - x_1^2 + x_1 x_2 - 7x_2$

(f) Maximize $f(X) = 7x_1 + 3x_1 x_2 - x_2$

3. Minimize $f(X) = x_1^2 + x_2^2$
   Subject to $x_1 - x_2 - 5 = 0$

4. Minimize $f(X) = x_1^2 + x_2^2 - 4x_1 - x_2$
   Subject to $x_1 + x_2 \geqslant 0$

5. Maximize $f(X) = 5x_1^2 + x_2^2 + 4$
   Subject to $4x_1 + x_2 \leqslant 4$
   $\qquad\qquad\quad 2x_1 + x_2 < 3$

6. Maximize $f(X) = 3x_1 + 2x_2$
   Subject to $x_1 + 2x_2 \leqslant 6$
   $\qquad\qquad\quad x_1 + x_2 \leqslant 4$

7. Minimize $f(X) = 5x_1^2 + 2x_2 - x_1 x_2$
   Subject to $x_1 + x_2 = 3$

8. Minimize $f(X) = x_1^2 + x_2^2$
   Subject to $x_1 + 3x_2 \geqslant 3$

3

SYSTEMS OF LINEAR EQUATIONS AND INEQUATIONS

INTRODUCTION

Linear relationships among variables are often suitable approximations of many real-world situations. For example, the labor (in standard hours) required to produce 400 units of item ABC may be roughly twice that required to produce 200 units. Or the costs of transporting materials a certain distance could be a linear function of the weight of the materials. Techniques for solving linear problems are relatively simple when compared with the techniques for solving nonlinear problems. For these reasons, we shall direct our attention in this chapter to several important concepts of linear algebra that underpin the Simplex Method of solving linear programming problems (to be discussed in Chapter 5). In a nutshell these concepts involve the graphical interpretation and solution of simultaneous linear equations and inequations.

GEOMETRY

OF LINEAR

EQUALITIES

An important problem in mathematics is determining the simultaneous solution to a set of linear equations. A simultaneous solution to two or more linear equations consists of those points common to all the equations. For example consider two straight lines, each of which is expressed in two variables:

$$2x_1 + 4x_2 = 8 \qquad \text{(Eq. 1)}$$

$$x_1 - 3x_2 = 6 \qquad \text{(Eq. 2)}$$

A plot of these two lines is shown in Figure 3-1. It can be seen that there is one point of intersection. Thus the simultaneous solution of the two lines results in a single point in the $x_1 - x_2$ plane. By solving for x_1 in the first equation and substituting this into the second equation, the value of x_2 at the point of intersection can be determined:

$$x_1 = \frac{8 - 4x_2}{2} \qquad \text{(from Eq. 1)}$$

$$\left(\frac{8 - 4x_2}{2}\right) - 3x_2 = 6 \qquad \text{(from Eq. 2)}$$

$$\text{or } x_2 = -\frac{2}{5}$$

Now the value of x_1 can be found by substituting $x_2 = -2/5$ into (Eq. 1): $2x_1 + 4 (-2/5) = 8$, or $x_1 = 24/5$. Thus the (x_1, x_2) coordinates of the intersection are $(24/5, -2/5)$.

When dealing with two straight lines it is possible for their simultaneous solution to be (a) a single point, (b) no points or (c) infinitely many points. The first situation (one point of intersection) was demonstrated in Figure 3-1, and the other two possibilities are shown in Figures 3-2 and 3-3. If there are three, four, five, etc., equations expressed in two variables (x_1 and x_2), we would plot them in the $x_1 - x_2$ plane to determine their common points of intersection (if any). Again, we would observe a single point, infinitely many points or no points of intersection.

Up to this point only equations in two variables have been discussed. Let us now consider linear equations having three variables. Such an equation can be represented in three-dimensional space, and it corresponds to a linear plane. For

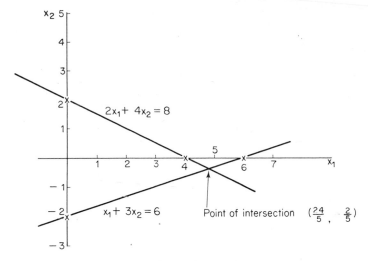

FIGURE 3-1   A Single Point of Intersection

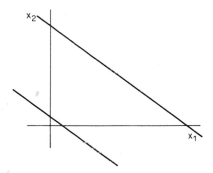

FIGURE 3-2   Two Parallel Lines (Points in Common = 0)

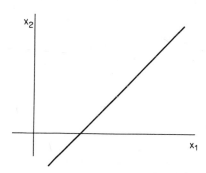

FIGURE 3-3   Two Coincident Lines (Points in Common = ∞)

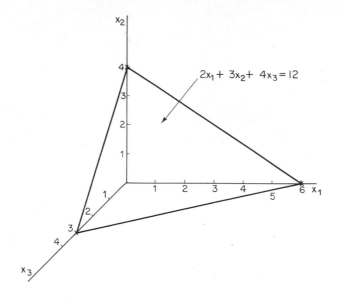

FIGURE 3-4   A Linear Plane Shown in Three Dimensions

instance, the plane represented by $2x_1 + 3x_2 + 4x_3 = 12$ is shown in Figure 3-4 for positive values of x_1, x_2 and x_3.

If we had three linear equations expressed in three variables, a simultaneous solution would be a single point or no points or infinitely many points. For equations with four or more variables these same three mutually exclusive solution sets exist. In linear programming problems the number of variables in each linear equation can become quite large, and determining a simultaneous solution is a time-consuming process. Often a computer is the only feasible means of determining a solution. Later in this chapter we shall investigate several methods for solving sets of equations with at most three variables.

GEOMETRY

OF LINEAR

INEQUALITIES

Because linear inequalities are present in the initial formulation of most linear programming problems, we shall briefly discuss the geometry of inequalities. A typical inequation is the following: $3x_1 + 4x_2 \leqslant 24$. The points satisfying this inequality are shown as a shaded region in the graph in Figure 3-5.

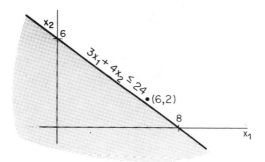

FIGURE 3-5   A Linear Inequality in Two Dimensions

From the graph it is apparent there are an infinite number of points (an area) satisfying the inequality. To see if a particular point, say (6,2), satisfies the inequality, we could plot the point in Figure 3-5 and note whether it lies in the shaded region. Or we could substitute the points $x_1 = 6$ and $x_2 = 2$ into the inequation as follows

$$3(6) + 4(2) \leqslant (?) \, 24$$

Because 26 is greater than 24, the point in question does not satisfy $3x_1 + 4x_2 \leqslant 24$.

To carry the example above one step further, suppose we consider two other inequalities in two variables:

$$x_1 + x_2 \geqslant 0$$

$$3x_1 - 2x_2 \geqslant -6$$

When all three inequalities are drawn in Figure 3-6 we can immediately identify the region that satisfies all of them simultaneously. This area is often referred to as a "feasible region" since it allows all inequalities to be met. In a later chapter we will be dealing with the feasible regions for inequalities expressed in terms of three or more variables. Accurate appraisals of feasible regions are essential to correct formulation of linear programming problems.

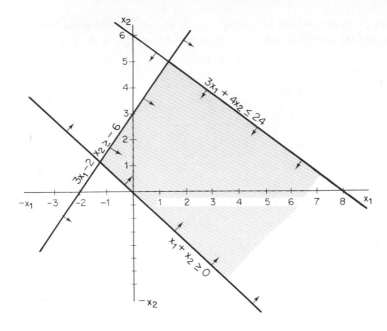

FIGURE 3-6   Three Inequalities Shown in Two Dimensions

LINEAR

DEPENDENCE AND

INDEPENDENCE

The procedures required to determine the determinant and inverse of a square matrix were described in Chapter 1. A set of n simultaneous linear equations in n variables has a unique (single) solution if the equation set has an inverse (a nonzero determinant). Furthermore, the existence of a nonzero determinant tells us that the rows (and columns) of the equation set are linearly independent. Before these two important concepts can be used in our later work with linear programming, we must be familiar with linear combinations of vectors and linear independence among vectors.

linear
combinations

If there are n column vectors, A_1, A_2, - - -, A_n, each with m components, then B is a linear combination of these n vectors when there are numbers c_1, c_2, - - -, c_n

such that $c_1 \cdot A_1 + c_2 \cdot A_2 + \cdots + c_n \cdot A_n = B$. For example, if $B = 4A_1 - 2A_2 - A_3$, then B is a linear combination of A_1, A_2 and A_3.

To illustrate further, let us express B as a linear combination of A_1 and A_2 by finding numbers c_1 and c_2 for which this relationship holds true:

$$\begin{bmatrix} 4 \\ 2 \end{bmatrix} = \begin{bmatrix} 6 \\ -1 \end{bmatrix} c_1 + \begin{bmatrix} 5 \\ 9 \end{bmatrix} c_2$$

where

$$B = \begin{bmatrix} 4 \\ 2 \end{bmatrix} \qquad A_1 = \begin{bmatrix} 6 \\ -1 \end{bmatrix} \qquad A_2 = \begin{bmatrix} 5 \\ 9 \end{bmatrix}$$

To find c_1 and c_2, we would solve the following set of linear equations:

$$4 = 6c_1 + 5c_2$$
$$2 = -c_1 + 9c_2$$

The solution to this set of equations yields $c_1 = 26/59$ and $c_2 = 16/59$. Thus,

$$\begin{bmatrix} 4 \\ 2 \end{bmatrix} = \begin{bmatrix} 6 \\ -1 \end{bmatrix} \frac{26}{59} + \begin{bmatrix} 5 \\ 9 \end{bmatrix} \frac{16}{59}$$

or

$$B = \frac{26}{59} A_1 + \frac{16}{59} A_2$$

By expressing a row or column vector (vector B) as a linear combination of A_1, A_2, ---, A_n, this is identical to solving a set of linear simultaneous equations, $A \cdot X = B$. It is also true that if a set of linear equations can be solved, we can express the vector B as a linear combination of row or column vectors comprising the particular set of equations under consideration.

linear independence

An n × n system of linear equations, $AX = B$, has the following coefficient matrix:

$$A = \begin{bmatrix} a_{11} & a_{12} & ---- & a_{1n} \\ a_{21} & a_{22} & ---- & a_{2n} \\ \cdot & \cdot & \cdot & \cdot \\ \cdot & \cdot & \cdot & \cdot \\ \cdot & \cdot & \cdot & \cdot \\ a_{n1} & a_{n2} & ---- & a_{nn} \end{bmatrix}$$

The column vectors of A are

$$A_1 = \begin{bmatrix} a_{11} \\ a_{21} \\ \cdot \\ \cdot \\ \cdot \\ a_{n1} \end{bmatrix}, \; A_2 = \begin{bmatrix} a_{12} \\ a_{22} \\ \cdot \\ \cdot \\ \cdot \\ a_{n2} \end{bmatrix}, \ldots, A_n = \begin{bmatrix} a_{1n} \\ a_{2n} \\ \cdot \\ \cdot \\ \cdot \\ a_{nn} \end{bmatrix}$$

The row vectors of A are

$$A_1 = [a_{11} \; a_{12} \; --- a_{1n}]$$
$$A_2 = [a_{21} \; a_{22} \; --- a_{2n}]$$
$$\cdot$$
$$\cdot$$
$$\cdot$$
$$A_n = [a_{n1} \; a_{n2} \; --- a_{nn}]$$

If all row vectors and column vectors are not independent the determinant of A will equal zero, det. $A = 0$, and the matrix is said to be singular. It can be seen from the adjoint method of calculating matrix inverses (Chapter 1) that a singular matrix will not have an inverse. Conversely, if a square matrix has an inverse and thus a unique solution, the row and column vectors of the coefficient matrix must be linearly independent. It then follows that if linear independence exists, the determinant of A will be nonzero.

With the above introductory remarks in mind, it is possible to utilize a simple definition for checking linear *dependence*. If we have a set of vectors A_1,

A_2, ---, A_n, we can discover whether they are linearly dependent by determining coefficients c_1, c_2, ---, c_n, not all zero, so that this linear combination holds true:

$$c_1 A_1 + c_2 A_2 + \cdots + c_n A_n = 0$$

In other words if at least one of the vectors in A can be expressed as a linear combination of the other vectors, the vector set is linearly dependent. We can also say that if all c_i must be zero to satisfy the above definition, the vectors are linearly independent. Thus, vectors are linearly independent when no one vector can be written as a linear combination of the others.

Example:

Suppose we have the following equations and would like to know whether they are linearly independent. (Square coefficient matrices such as the one below are predominant in our later work with linear programming.)

$$4x_1 + 3x_2 \qquad = 5$$
$$2x_1 + x_2 + x_3 = 6$$
$$x_1 + 2x_2 + 2x_3 = 3$$

If we let

$$A = \begin{bmatrix} 4 & 3 & 0 \\ 2 & 1 & 1 \\ 1 & 2 & 2 \end{bmatrix}, \quad X = \begin{bmatrix} x_1 \\ x_2 \\ x_3 \end{bmatrix} \text{ and } B = \begin{bmatrix} 5 \\ 6 \\ 3 \end{bmatrix}$$

then $AX = B$. We will first work with column vectors in A to see if they are linearly dependent. If they are, then

$$c_1 A_1 + c_2 A_2 + c_3 A_3 = 0 \qquad (\text{All } c_i \neq 0)$$

Written out, this equation becomes:

$$c_1 \begin{bmatrix} 4 \\ 2 \\ 1 \end{bmatrix} + c_2 \begin{bmatrix} 3 \\ 1 \\ 2 \end{bmatrix} + c_3 \begin{bmatrix} 0 \\ 1 \\ 2 \end{bmatrix} = \begin{bmatrix} 0 \\ 0 \\ 0 \end{bmatrix}$$

These three simultaneous equations result.

$$4c_1 + 3c_2 \qquad\quad = 0$$
$$2c_1 + c_2 + c_3 = 0$$
$$c_1 + 2c_2 + 2c_3 = 0$$

From the first equation, $c_1 = -3/4 \cdot c_2$. Inserting this value of c_1 into the second and third equations yields the following:

$$
\left.
\begin{aligned}
(-3/2)c_2 + c_2 &= -c_3 \\
(-3/4)c_2 + 2c_2 &= -2c_3
\end{aligned}
\right\}
\quad \text{or} \quad
\begin{aligned}
-0.5c_2 &= -c_3 \\
1.25c_2 &= -2c_3
\end{aligned}
$$

We have contradictory results since $c_2 = 2c_3$ and also $c_2 = (-8/5)c_3$. This establishes that we have linearly independent equations. If the substitution of the expression for c_1 into the second and third equations had produced identical (noncontradictory) results, linear dependency would have existed among column vectors.

We can also use row vectors to check for linear dependence. When

$$
A = \begin{bmatrix} A_1 \\ A_2 \\ A_3 \end{bmatrix} = \begin{bmatrix} 4 & 3 & 0 \\ 2 & 1 & 1 \\ 1 & 2 & 2 \end{bmatrix}
$$

does $c_1 A_1 + c_2 A_2 + c_3 A_3 = 0$? To answer this question, we must again evaluate the c_i coefficients:

$$c_1 [4\ 3\ 0] + c_2 [2\ 1\ 1] + c_3 [1\ 2\ 2] = [0\ 0\ 0]$$
$$4c_1 + 2c_2 + c_3 = 0$$
$$3c_1 + c_2 + 2c_3 = 0$$
$$\phantom{3c_1 + {}}c_2 + 2c_3 = 0$$

We discover that $c_1 = 0$, which means that $c_3 = 0$ and $c_2 = 0$. When all coefficients are zero, we have a system of linearly independent equations.

Because rows (and columns) were independent, we could conclude that A has an inverse (i.e. A is nonsingular). In checking the validity of this statement we find that

$$\text{Det. A} = -9$$

$$A^{-1} = \frac{\text{Adjoint}}{\text{Det. A}} = \begin{bmatrix} 0 & 2/3 & -1/3 \\ 1/3 & -8/9 & 4/9 \\ -1/3 & 5/9 & 2/9 \end{bmatrix}$$

and $A^{-1} A = I$.

To summarize, if it can be demonstrated that contradictory values of c_i exist, or when solutions to simultaneous equations in c_i can only be found when all $c_i = 0$, a square matrix consists of linearly independent equations. Linear independence among rows in a square matrix implies that the columns are linearly independent and vice versa. In linear programming problems to be discussed later, it is essential that the column vectors of the coefficient matrix be linearly independent.

SOLUTIONS OF

SIMULTANEOUS

LINEAR

EQUATIONS

In linear programming problems we will be dealing with sets of linear equalities in which the number of variables, n, is often large and greater than or equal to the number of equations, m, (i.e. $n \geqslant m$). Because there are numerous solutions to linear equations for which $n \geqslant m$, we must impose other conditions on the problem so that unique values of the variables can be determined. Such conditions are fully discussed in a later chapter, and our intent here is to demonstrate several procedures for solving simultaneous linear equations. One of these, the Gauss-Jordan method, is particularly useful when dealing with sets of linear equations having three or more variables, and it is easily programmed for the computer.

Problems with a maximum size of $n = m = 3$ are dealt with in this chapter. Principal computational features of the various methods can be adequately illustrated in matrices of this size.

There are three general cases to consider in solving simultaneous linear equations: (1) $m > n$, (2) $m = n$ and (3) $m < n$. Even though we will focus our attention on the third case, the first two cases are treated briefly below.

**more linear
equations than
unknown quantities
(m > n)**

When there are m simultaneous linear equations expressed in n variables and m > n, there can be zero, one or infinitely many solutions. Frequently, the m equations do not intersect at a common point and there are no solutions to the problem. If the m equations intersect at a single point, we would have a single solution. Infinitely many solutions could occur when all m equations correspond to the same line (the equations are coincident with one another).

To illustrate a situation in which m > n and one solution exists, consider the equations in Fig. 3-7. To facilitate the use of matrix algebra, variables continue to be represented by subscripted lower-case letters.

The solution to this set of linear equations can be determined from any two of the three equations since we have more equations than unknowns. The third equation can then be used to confirm the existence of a single solution. From the first two equations we find $x_1 = 2$ and $x_2 = -1$. The third equation confirms that $(2, -1)$ is the solution to this problem since $2 - 3(-1) = 5$. If the third equation resulted in an inequality when values of $x_1 = 2$ and $x_2 = -1$ were substituted into it, we would have reason to believe that no solutions existed. Of course if all three equations represented the same line, no unique values of x_1 and x_2 could be determined and there would be an infinite number of solutions to the equation set.

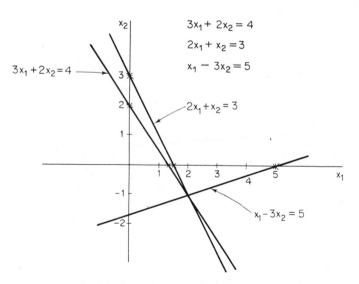

$$3x_1 + 2x_2 = 4$$
$$2x_1 + x_2 = 3$$
$$x_1 - 3x_2 = 5$$

FIGURE 3-7   Three Equations, Two Unknowns with a Unique Solution

equal number of
linear equations and
unknown quantities
(m = n)

In the case of a square coefficient matrix, a solution can be found (when a single solution exists) by several methods. When the determinant of the coefficient matrix is nonzero, we can say that a single solution exists and that the linear equations are independent. If we discover a zero determinant, there can either be no solution to the equations or an infinite number of solutions. (Recall that linear dependence exists among the equations when det. A = 0.) In this section we demonstrate (1) the matrix inversion method and (2) Cramer's rule for solving an n × n system of linear equations when the equation set is nonsingular (det. A ≠ 0). The Gauss-Jordan procedure can also be used; however, it is of special interest when n > m and will be thoroughly treated in the next section.

1. *Matrix Inversion*. Consider this set of simultaneous linear equations (m = n = 3):

$$3x_1 + 2x_2 + x_3 = 9$$

$$x_1 - x_2 + 3x_3 = 6$$

$$-x_1 - x_2 + x_3 = 3$$

$$\text{Let } A = [A_1 \ A_2 \ A_3] = \begin{bmatrix} 3 & 2 & 1 \\ 1 & -1 & 3 \\ -1 & -1 & 1 \end{bmatrix}, \ X = \begin{bmatrix} x_1 \\ x_2 \\ x_3 \end{bmatrix} \text{ and } B = \begin{bmatrix} 9 \\ 6 \\ 3 \end{bmatrix}$$

so that AX = B. To solve for x_1, x_2 and x_3, we could determine A^{-1} and multiply by B; i.e., $X = A^{-1}B$. The number of columns in A^{-1} equals the number of rows in B so matrix multiplication (i.e., $A^{-1}B$) can be performed. As you recall, to compute A^{-1} it is necessary that A be nonsingular and a square matrix. (A procedure for computing the inverse of a matrix was discussed previously in Chapter 1.) Because A is a 3 by 3 matrix, det. A is calculated as follows, by expanding along the first row:

$$\text{det. A} = (-1)^{1+1}(3 \cdot 2) + (-1)^{1+2}(2 \cdot 4) + (-1)^{1+3}(1 \cdot -2)$$

$$= 6 - 8 - 2$$

$$= -4$$

Therefore, A is nonsingular. Next we will invert A by using the adjoint and the determinant. The adjoint (A_{adj}.) equals the transpose of cofactors of a_{ij} in A. That is,

$$A^{-1} = \frac{1}{\det.A} \cdot A_{adj} \quad \text{and} \quad A_{adj} = |C_{ij}|^T$$

Performing these operations yields the adjoint matrix:

$$A_{adj} = \begin{bmatrix} +\begin{vmatrix} -1 & 3 \\ -1 & 1 \end{vmatrix} & -\begin{vmatrix} 2 & 1 \\ -1 & 1 \end{vmatrix} & +\begin{vmatrix} 2 & 1 \\ -1 & 3 \end{vmatrix} \\ -\begin{vmatrix} 1 & 3 \\ -1 & 1 \end{vmatrix} & +\begin{vmatrix} 3 & 1 \\ -1 & 1 \end{vmatrix} & -\begin{vmatrix} 3 & 1 \\ 1 & 3 \end{vmatrix} \\ +\begin{vmatrix} 1 & -1 \\ -1 & -1 \end{vmatrix} & -\begin{vmatrix} 3 & 2 \\ -1 & -1 \end{vmatrix} & +\begin{vmatrix} 3 & 2 \\ 1 & -1 \end{vmatrix} \end{bmatrix} = \begin{bmatrix} 2 & -3 & 7 \\ -4 & 4 & -8 \\ -2 & 1 & -5 \end{bmatrix}$$

Finally,

$$A^{-1} = \frac{A_{adj}}{\det.A} = -\frac{1}{4} \begin{bmatrix} 2 & -3 & 7 \\ -4 & 4 & -8 \\ -2 & 1 & -5 \end{bmatrix} = \begin{bmatrix} -\frac{1}{2} & \frac{3}{4} & -\frac{7}{4} \\ 1 & -1 & 2 \\ \frac{1}{2} & -\frac{1}{4} & \frac{5}{4} \end{bmatrix}$$

The above operations can be checked by showing that $A \cdot A^{-1} = I$. It is now possible to solve for $X = A^{-1}B$ as shown below:

$$\begin{bmatrix} x_1 \\ x_2 \\ x_3 \end{bmatrix} = \overset{3 \times 3}{\begin{bmatrix} -\frac{1}{2} & \frac{3}{4} & -\frac{7}{4} \\ 1 & -1 & 2 \\ \frac{1}{2} & -\frac{1}{4} & \frac{5}{4} \end{bmatrix}} \overset{3 \times 1}{\begin{bmatrix} 9 \\ 6 \\ 3 \end{bmatrix}} = \overset{3 \times 1}{\begin{bmatrix} -\frac{21}{4} \\ 9 \\ \frac{27}{4} \end{bmatrix}}$$

or $x_1 = -21/4$, $x_2 = 9$, $x_3 = 27/4$.

2. *Cramer's Rule.* This is a procedure to solve n linear equations with n unknown quantities when det. A $\neq$ 0. Here we are dealing with a system of equations of the form AX = B. To solve for a particular x_j ($1 \leq j \leq n$), we first substitute the column vector B for the jth column vector in the coefficient matrix A. Next we solve for x_j in this manner:

$$x_j = \frac{\det.A_j}{\det.A}$$

where A_j is the matrix that results from replacing the jth column of A by the column vector B.

To demonstrate Cramer's rule, we will again solve this set of linear equations:

$$3x_1 + 2x_2 + x_3 = 9$$

$$x_1 - x_2 + 3x_3 = 6$$

$$-x_1 - x_2 + x_3 = 3$$

The values of x_1, x_2 and x_3 are determined as follows by using the above definition. (Recall that det.A = -4.)

$$x_1 = \frac{\begin{bmatrix} \overset{B}{\underset{\downarrow}{9}} & \overset{A_2}{\underset{\downarrow}{2}} & \overset{A_3}{\underset{\downarrow}{1}} \\ 6 & -1 & 3 \\ 3 & -1 & 1 \end{bmatrix}}{\det.A} = \frac{9(-1+3)-6(2+1)+3(6+1)}{-4} = -\frac{21}{4}$$

$$x_2 = \frac{\begin{bmatrix} \overset{A_1}{\underset{\downarrow}{3}} & \overset{B}{\underset{\downarrow}{9}} & \overset{A_3}{\underset{\downarrow}{1}} \\ 1 & 6 & 3 \\ -1 & 3 & 1 \end{bmatrix}}{\det.A} = \frac{3(6-9)-1(9-3)-1(27-6)}{-4} = 9$$

$$x_3 = \frac{\begin{bmatrix} \overset{A_1}{\underset{\downarrow}{3}} & \overset{A_2}{\underset{\downarrow}{2}} & \overset{B}{\underset{\downarrow}{9}} \\ 1 & -1 & 6 \\ -1 & -1 & 3 \end{bmatrix}}{\det.A} = \frac{3(-3+6)-1(6+9)-1(12+9)}{-4} = \frac{27}{4}$$

As you see, both methods provide identical results. When the number of variables exceeds three, matrix inversion and Cramer's rule become computationally cumbersome in solving problems where m = n. The Gauss-Jordan method is better suited to larger problems where m = n, and it is especially useful where m < n.

**fewer linear
equations than
unknown quantities
(m < n)**

When a set of linear equations has more unknown quantities (variables) than equations, it is no longer possible to find a single, unique solution. Here we use the Gauss-Jordan procedure in solving problems of this nature and again note that the Simplex Method of linear programming utilizes it.

To illustrate the difficulties of solving systems of linear equations for which n > m, consider these two equations:

$$2x_1 - 4x_2 - x_3 = 6$$
$$-x_1 + 3x_2 + 2x_3 = 12$$

In attempting to determine values of x_1, x_2 and x_3, it becomes apparent there are a large number of solutions since we have more variables than equations. To simplify the problem somewhat, suppose that x_3 cannot exceed 10. If we let $x_3 = 10$, the equations above become:

$$\left.\begin{array}{l} 2x_1 - 4x_2 - 10 = 6 \\ \\ -x_1 + 3x_2 + 20 = 12 \end{array}\right\} \quad \text{or} \quad \begin{array}{l} 2x_1 - 4x_2 = 16 \\ \\ -x_1 + 3x_2 = -8 \end{array}$$

By solving for x_1 and x_2 we find that $x_1 = 8$ and $x_2 = 0$.

Similarly, when $x_3 = 9$ the values of x_1 and x_2 become 10.5 and 1.5, respectively. After repeating this procedure for other values of $x_3 < 10$, numerous combinations of (x_1, x_2) would be permissible solutions to the problem. When x_1 or x_2 have constraints placed on them, it is clear that sets of linear equations with n > m may have solutions that are not trivial and the original equation set usually has an infinitely large number of solutions.

The general procedure followed in solving linear equations with n > m is to arbitrarily assign zero values to x_{m+1}, x_{m+2}, ..., x_n (i.e. n−m of the variables) and then determine values for the remaining m × m system of equations by the Gauss-Jordan method. Other methods, such as matrix inversion or Cramer's rule, could also be employed, but they are usually more time consuming in larger problems. Once a solution to the m × m system of equations has been determined (assuming, of course, a nonzero det. A), it is called a basic solution if there are no more than m nonzero variables. You can see there are many basic solutions to a set of linear equations when n > m. More will be said about basic solutions and their relationship to linear programming in Chapter 5.

The Gauss-Jordan Method.   This method is normally utilized to solve large systems of linear equations (m = n or m < n). Let us illustrate how the Gauss-Jordan method works by using our previous two equations:

$$2x_1 - 4x_2 - x_3 = 6$$

$$-x_1 + 3x_2 + 2x_3 = 12$$

Suppose for the moment that $x_3 = 0$, so that

$$2x_1 - 4x_2 = 6 \qquad \text{(Eq. 3)}$$

$$-x_1 + 3x_2 = 12 \qquad \text{(Eq. 4)}$$

Now let us eliminate x_1 from (Eq. 4). If we divide (Eq. 3) by 2, we obtain $x_1 - 2x_2 = 3$ (Eq. 3a). Thus, the first step is to divide the first equation in the equation set by the coefficient of x_1. To eliminate x_1 in (Eq. 4) we multiply (Eq. 3a) by the negative of the coefficient of x_1 in (Eq. 4). To this point, the results are as follows.

$$-x_1 + 3x_2 = 12 \qquad \text{(Eq. 4)}$$

$$x_1 - 2x_2 = 3 \qquad \text{(Eq. 3a × +1)}$$

We can now eliminate $-x_1$ in (Eq. 4) by replacing (Eq. 4) by the *sum* of (Eq. 4) and (Eq. 3a × 1) which is

$$0 + x_2 = 15$$

Therefore $x_2 = 15$ and from (Eq. 3a × +1), $x_1 = 3 + 2x_2 = 33$.

We could have obtained this same result by eliminating x_2 from (Eq. 3). First, divide (Eq. 4) by the coefficient of x_2. This becomes

$$-\frac{x_1}{3} + x_2 = 4 \qquad\qquad\text{(Eq. 4a)}$$

We next multiply (Eq. 4a) by the negative of the coefficient of x_2 in (Eq. 3).

$$+4\left[\frac{-x_1}{3} + x_2 = 4\right] = \frac{-4x_1}{3} + 4x_2 = 16 \qquad\text{(Eq. 4a × 4)}$$

Next eliminate $-4x_2$ in (Eq. 3) by replacing (Eq. 3) with the sum of (Eq. 4a × 4) and (Eq. 3). The result is $\frac{2}{3} x_1 + 0 = 22$. Again we see that $x_1 = 33$ and $x_2 = 15$.

With the same original equations, suppose now $x_1 = 0$ so that

$$-4x_2 - x_3 = 6$$

$$3x_2 + 2x_3 = 12$$

The reader should be able to use the Gauss-Jordan method to show that $x_2 = -4.8$ and $x_3 = 13.2$.

To demonstrate further the Gauss-Jordan method, consider a set of linear equations with $m = n = 3$.

$$x_1 - x_2 + x_3 = -2 \qquad\qquad\text{(Eq. 5)}$$

$$2x_1 - 4x_2 + 3x_3 = -3 \qquad\qquad\text{(Eq. 6)}$$

$$0 - 2x_2 + 3x_3 = 7 \qquad\qquad\text{(Eq. 7)}$$

These equations could be written in the following form.

$$\begin{bmatrix} 1 & -1 & 1 \\ 2 & -4 & 3 \\ 0 & -2 & 3 \end{bmatrix} \begin{bmatrix} x_1 \\ x_2 \\ x_3 \end{bmatrix} = \begin{bmatrix} -2 \\ -3 \\ 7 \end{bmatrix}$$

We can use the Gauss-Jordan method to determine X such that the left side of the equation equals the right side. To eliminate x_1 in the second equation, we can subtract twice (Eq. 5) from (Eq. 6), i.e. (Eq. 6a = Eq. 6 - 2 Eq. 5). The result is

$$0 - 2x_2 + x_3 = 1 \qquad \text{(Eq. 6a)}$$

If (Eq. 6a) is multiplied by -1 and added to (Eq. 7), we obtain (Eq. 7a = Eq. 7 - Eq. 6a):

$$0 + 0 + 2x_3 = 6 \qquad \text{(Eq. 7a)}$$

Our final set of equations is now in triangular form, which allows values of x_1, x_2 and x_3 to be quickly determined:

(Eq. 5)	x_1	$-$	x_2	$+$	x_3	$=$	-2		$x_1 = -4$
(Eq. 6a)	0	$-$	$2x_2$	$+$	x_3	$=$	1	or	$x_2 = 1$
(Eq. 7a)	0	$+$	0	$+$	$2x_3$	$=$	6		$x_3 = 3$

Another set of linear equations typical of many linear programming problems appears below:

$$2x_1 + 4x_2 + x_3 + x_4 = 8 \qquad \text{(Eq. 8)}$$

$$x_1 + 5x_2 \qquad + 3x_4 = 8 \qquad \text{(Eq. 9)}$$

A physical situation giving rise to this set of linear equations could easily involve the production schedule of a highly specialized machine shop. Suppose x_1, x_2, x_3 and x_4 represent four different items that must be routed through various machining operations. The first equation above expresses the time (in hours) required to complete the first operation for item 1, item 2, item 3 and item 4, respectively. Similarly, the second equation indicates time requirements for the four items in the second operation. Finally, these equations tell us that total time available each day for production in operations 1 and 2 is eight hours. (Restrictions of this nature could be imposed, for example, by the availability of highly skilled machinists.)

No information has been given regarding costs or profits of each item produced. Thus, our problem reduces to one of deciding how many units of each item to produce in view of limitations on machine operation time.

To solve these equations, $n - m = 2$ of the variables must be set equal to zero. Suppose that $x_2 = x_3 = 0$. We can use the Gauss-Jordan method on the entire equation set in determining values of x_1 and x_4. First divide (Eq. 8) by the coefficient of x_1:

$$x_1 + 2x_2 + \frac{x_3}{2} + \frac{x_4}{2} = 4 \qquad \text{(Eq. 8a)}$$

Then multiply (Eq. 9) by -1, the negative of the x_1 coefficient in (Eq. 9), and replace (Eq. 9) by the sum of (Eq. 8a) and (Eq. 9 $\times$ -1). The result is

$$-3x_2 + \frac{x_3}{2} - \frac{5x_4}{2} = -4$$

Since x_2 and x_3 are zero, we see that $x_4 = 8/5$. Then it follows that $x_1 = 16/5$.

Instead of removing x_1 (Eq. 8), we could have eliminated x_4 from (Eq. 9) and then determined x_1. Begin with dividing (Eq. 9) by the coefficient of x_4 (Eq. 9a = Eq. 9 $\div$ 3):

$$\frac{x_1}{3} + \frac{5x_2}{3} + x_4 = \frac{8}{3} \qquad \text{(Eq. 9a)}$$

Then multiply (Eq. 8) by -1 and replace (Eq. 8) by the sum of (Eq. 9a) and (Eq. 8 $\times$ -1), i.e. (Eq. 8a = Eq. 9a $-$ Eq. 8):

$$\frac{-5}{3} x_1 - \frac{7}{3} x_2 - x_3 = \frac{-16}{3}$$

Thus, again we see that $x_1 = 16/5$ and $x_4 = 8/5$.

The same procedure would be repeated to solve for the values of any pair of variables. Because it is implied that our production units are integer quantities, we would be tempted to decide that $x_1 = 3$ units and $x_4 = 2$ units in the previous example. More will be said in Chapter 6 regarding how to ensure integer-valued variables in linear programming problems.

To summarize the Gauss-Jordan method, several different operations can be performed without affecting the system of linear equations being studied. We can:

1. interchange any two equations,

2. multiply both sides of an equation by a -1 (or any nonzero constant),

3. add or subtract multiples of any equation to or from another equation.

These operations should be applied to a set of equations until they are reduced to triangular form so that desired values can be quickly ascertained.

Thus far our example problems have had a single solution. How can we determine whether there is no solution or infinitely many solutions when using the Gauss-Jordan method? In the last row operation we found an equation of the form, $a_{mn}x_n = b_n$. For a nonzero value of b_n there will be no value of x_n that satisfies the above equation if $a_{mn} = 0$. However, if $b_n = 0$ and $a_{mn} = 0$ there are infinitely many values of x_n for which the equation, $0 \cdot x_n = 0$, holds true. Either situation (i.e. no solutions or infinitely many) could have been discovered by checking the determinant of the coefficient matrix. If det. $A = 0$, rows or columns of the coefficient matrix are linearly dependent and hence no unique solution exists.

SUGGESTED ADDITIONAL READINGS

Bowen, Earl K. *Mathematics with Applications in Management and Economics.* Homewood, Ill.: Richard D. Irwin, Inc. 1967.

Campbell, Hugh G. *An Introduction to Matrices, Vectors and Linear Programming.* New York: Appleton-Century-Crofts, 1965.

Chung, A. *Linear Programming.* Columbus, Ohio: Merrill Books, Inc., 1963.

Hadley, G. *Linear Algebra.* Reading, Mass.: Addison-Wesley, 1961.

EXERCISES

1. Determine if the following sets of equations are linearly independent or dependent:

   (a) $2x_1 + 3x_2 = 8$
   $5x_1 + x_2 = 7$

   (b) $3x_1 + x_2 + x_3 = 4$
   $5x_1 + 7x_2 + 13x_3 = 18$
   $x_1 + x_2 + 3x_3 = 4$

(c) $7x_1 + 3x_2 - 14x_3 = -4$
$x_1 + 2x_2 - x_3 = 2$
$x_1 - x_2 + 7x_3 = 6$

(d) $14x_1 + x_2 = 3$
$28x_1 + 2x_2 = 6$

(e) $7x_1 + x_2 + 2x_3 = 10$
$3x_2 + 2x_2 + 4x_2 = 9$
$2x_1 + x_3 + 2x_3 = 5$

2. Solve the following sets of equations by using matrix inversion, Cramer's rule and the Gauss-Jordan procedure (if possible):

(a) $x_1 + x_2 = 2$
$3x_1 + x_2 = 4$

(b) $2x_1 + x_2 = 5$
$4x_1 + 2x_2 = 10$

(c) $3x_1 + x_2 + 4x_3 = 16$
$x_1 + x_2 + x_3 = 6$
$2x_1 + x_2 + x_3 = 8$

(d) $x_1 + 7x_2 + 3x_3 = 12$
$9x_1 + 4x_2 + 5x_3 = 18$
$x_1 + 2x_2 + 6x_3 = 6$

(e) $4x_1 + x_2 + 2x_3 = 4$
$7x_1 + \qquad + 4x_3 = 7$
$2x_1 + 9x_2 \qquad = 2$

Note: There may be one unique solution, multiple solutions or no solutions.

4

A GRAPHICAL INTRODUCTION TO LINEAR PROGRAMMING

INTRODUCTION

Many industrial problems involve the allocation of limited resources for the purpose of obtaining the best possible results from their use. "Best possible results" usually means that our aim is to maximize profits or to minimize costs. It also implies that several alternatives exist from which to choose to accomplish a specific goal. Thus, our problem is to allocate fixed and known amounts of resources in satisfying a given goal such that we maximize profits (or minimize costs) for feasible alternatives under consideration.

Linear programming is probably the best known and most widely used optimization technique for solving certain types of resource allocation problems. It adequately represents a wide variety of real-world problems and can be quickly encoded for solution with digital computers. The Simplex Method of linear programming was developed in 1947 by George Dantzig and made the solution of large problems computationally tractable. Today linear programming is routinely used by many industries including agriculture, steel, chemicals, airlines, petroleum and utilities.

Several conditions must be met before linear programming can become a reliable tool. First, we are concerned with specifying *non-negative* values of a set of variables that optimize a *linear* function expressed in terms of these variables. Second, the optimization of this function must also satisfy one or more linear constraints that mathematically take into account the availability of resources. Linearily implies, for example, that profit (or cost) per unit of output remains constant regardless of production level. Similarly, total resources consumed are assumed to be a linear function of the production level.

A RESOURCE ALLOCATION EXAMPLE THAT ILLUSTRATES THE BASIC CONCEPTS OF LINEAR PROGRAMMING

To illustrate this special type of resource allocation problem, consider the manufacture of Item a and Item b. Each unit produced requires a certain amount of machining time (i.e., standard time per operation) in each of three departments as shown below.

		Department		
		1	2	3
Item	a	40 min.	24 min.	20 min.
	b	30 min.	32 min.	24 min.

Each department works a standard day consisting of 480 minutes, so it is clear that with no overtime there is a definite limit to the availability of machining time in each department. Because we cannot produce negative amounts of Item a and Item b, nor can we utilize a negative amount of time in their manufacture, none of the factors present in this problem can be negative. Suppose further that the profit per unit of Item a and Item b is $5 and $8, respectively.

In our simple problem, the aim is to maximize profits. This can be expressed mathematically by the following equation

$$\text{Maximize } P = 5a + 8b$$

where a = the number of units of Item a manufactured, and b = the number of units of Item b produced. From a quick inspection it should be obvious that the above function is linear. This equation, known as the objective function, is expressed in terms of the *decision variables* (i.e., quantities that we can control so as to maximize profits). The *constraints* in this problem concern available machining time in each department and are also linear in terms of our two decision variables as seen below.

$$40a + 30b \leqslant 480 \qquad \text{(constraint on available time in department 1)}$$

$$24a + 32b \leqslant 480 \qquad \text{(constraint on available time in department 2)}$$

$$20a + 24b \leqslant 480 \qquad \text{(constraint on available time in department 3)}$$

If it is assumed that only two products are being manufactured and that all machining time in departments 1, 2 and 3 is available solely for this purpose, we can formulate this problem as a linear programming problem since the special characteristics involving linearity of the objective function and constraints are present here. Thus, the problem can be written:

$$\text{Maximize} \quad P = 5a + 8b$$
$$\text{Subject to} \quad 40a + 30b \leqslant 480$$
$$24a + 32b \leqslant 480$$
$$20a + 24b \leqslant 480$$
$$a \geqslant 0$$
$$b \geqslant 0$$

Our task now is to determine values of a and b that maximize profits and satisfy the linear constraints.

Numerous alternatives are available for the solution of this problem. Consider, for example, what would happen if we decided to produce only Item a *or* Item b.

Item a Only	*Item b only*
40 a ⩽ 480, or a ⩽ 12.0	30 b ⩽ 480, or b ⩽ 16.0
24 a ⩽ 480, or a ⩽ 20.0	32 b ⩽ 480, or b ⩽ 15.0
20 a ⩽ 480, or a ⩽ 24.0	24 b ⩽ 480, or b ⩽ 20.0
No more than 12 units can be produced without violating constraint on time of department 1.	No more than 15 units can be produced without violating constraint on time of department 2.
P = 12 units ($5/unit) = $60	P = 15 units ($8/unit) = $120

Other alternatives involving various combinations of Item a and Item b could also be proposed and evaluated. However, in larger problems enumerating all possible combinations of the decision variables could be quite time consuming. What we need is a systematic solution procedure that will seek out the optimum combination of Items a and b to be manufactured.

The Simplex Method, discussed in the next chapter, has been developed for this purpose. For the remainder of this chapter, though, we will investigate graphical procedures for dealing with linear programming problems having two decision variables. Through graphical means, it is hoped the reader will form a clear picture of how linear programming works. An extension of these notions to larger problems will then facilitate a thorough understanding of the Simplex Method and other linear programming algorithms to be discussed later.

Returning to our manufacturing problem, a solution can be discovered by first drawing a graph with units of Item a along the ordinate and units of Item b along the abscissa. If the constraint equations are then plotted on this graph, we would have Figure 4-1.

The shaded area, called the *feasible region*, defines a convex polygon that contains the optimal solution to this problem. The feasible region is *convex* because a straight line connecting any two points in it will lie entirely within the region (refer to Chapter 2 for a discussion of convexity).

As you can easily verify, the feasible region permits these constraints to be satisfied:

$$a \geqslant 0$$

$$b \geqslant 0$$

$$40a + 30b \leqslant 480$$

$$24a + 32b \leqslant 480$$

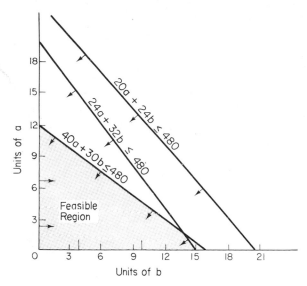

FIGURE 4-1 The Feasible Region

Thus, any point satisfying all the constraints and non-negativity conditions is a feasible solution to our problem. The last constraint (available time in department 3) does not lie in the shaded region because it is not a *binding* constraint. This means that there is no way of using all available time in department 3 to produce Item a and Item b without violating constraints on time in departments 1 and 2. If it were possible to schedule overtime operation in these two departments, the constraint on department 3 could become binding (or "active"), but this possibility is not being considered in the present problem. Thus, we do not regard the time available in department 3 as a limited resource and only two constraints are necessary.

Now that the feasible region for a solution has been defined, we must attempt to maximize our objective function by specifying the optimal number of units of Items a and b to manufacture. This is done by superimposing the objective function on Figure 4-1 and moving it as far as possible from the origin without leaving the feasible region. (We move away from the origin since we are trying to maximize profits.) This can be understood readily by referring to Figures 4-2 and 4-3.

In Figure 4-2 the objective function is plotted. When $5a + 8b = P$ where P is any constant value, there is a direct relationship between a and b that can be used to plot the objective function. For example, suppose $P = \$120$. When $a = 0$ and $b = 15$, the profit is $120. But when $a = 24$ and $b = 0$, profit will also be $120. Table 4-1 illustrates a few of the other combinations of a and b resulting in a profit of $120.

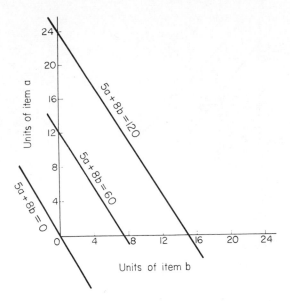

FIGURE 4-2   The Objective Function

TABLE 4-1
COMBINATIONS OF a AND b GIVING $120 PROFIT

a	b	Profit(P)
10	8.75	120
4.80	12	120
8	10	120
16	5	120

To plot the objective function, $120 = 5a + 8b$, we could solve for $a = 24 - (8/5)b$, which is a straight line with a slope of $-8/5$ and an a-intercept of 24. This line is plotted in Figure 4-2. Also shown are other members of a family of objective functions with slopes of $-8/5$ and a-intercepts of $P/5$ appearing as a series of parallel lines.

    Some values of P, however, will cause the decision variables to violate one or more constraints. That is, all or a part of the line formed by $5a + 8b = P$ may not lie in the feasible region defined by our constraint set. The combination of a and b that we want to determine is the one allowing the maximum value of P to occur in the equation $5a + 8b = P$ while this same equation lies in the feasible region at one or more points.

An example of too large a value of P is given in Figure 4-3, where P = 200. This graph also illustrates a situation in which the objective function lies in the feasible region but is not at its maximum possible value. In this case, the objective function, $5a + 8b = 60$, must move farther away from the origin to be maximized.

Finally, Figure 4-4 shows the maximum value of the objective function to be $5(0)+8(15)=120$ for $a = 0$ and $b = 15$. Note in Figure 4-4 that the objective function touches the feasible region at exactly one point. This is typical of most linear programming problems and will be explained later.

It is also clear that the feasible region depends only on the constraint equations and is unaffected by changes in the objective function. Moreover the optimal solution is determined by the slope of the objective function, given a certain set of constraints. For example, suppose the objective function were $P = 9a + 3b$. The optimal solution would then be $a = 12$, $b = 0$ with a profit of $108.

Care must be taken to formulate constraints correctly. Contradictory results will be obtained by erroneously reversing one or more of the inequalities in the constraint set. For example, suppose these constraints are specified.

$$a + b \leqslant 3$$

$$2a + 4b \geqslant 14$$

$$a, b \geqslant 0$$

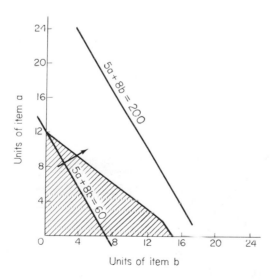

FIGURE 4-3   The Objective Function and the Feasible Region

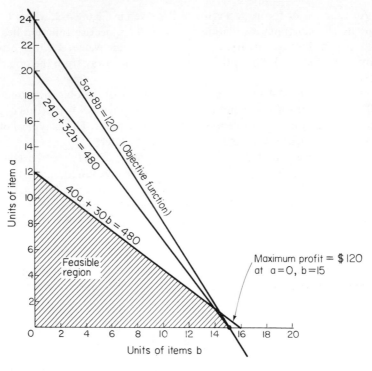

FIGURE 4-4 A Graphical Solution to the Manufacturing Problem

If we attempt to define the feasible region, we find that no point satisfies all constraints.

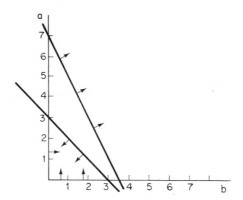

This shows that the feasible region is an empty set, and our problem must be modified before we can proceed. Linear programming problems can also fail to have solutions when the feasible region is unbounded. That is, we can maximize the objective function without limit while staying in the feasible region. To envision this difficulty, try to solve this problem graphically:

$$\text{Maximize} \quad P = 3a + 4b$$

$$\text{Subject to} \quad -a + b \leqslant 1$$

$$-a/2 + b \leqslant 3$$

$$a, b \geqslant 0$$

The optimal solution could be found to our illustrative production problem by making inequalities into equalities and then solving a simultaneous set of linear equations. In this case there are only two equations to deal with simultaneously since we have two decision variables. But when several constraints are imposed on the problem, it is often difficult to determine *which* two equations to solve to find the maximum (or minimum) value of the objective function. This raises another interesting possibility for solving linear programming problems in addition to the graphical method.

If we determine the points of intersection for every pair of constraint equations (equalities), it would be possible to find the point resulting in a maximum value of $P = 5a + 8b$. Suppose for the moment that we do not know constraint 3 is nonbinding in the problem. All pairs of constraint equations and their points of intersection are shown below. There are $\binom{5}{2} = 5!/3!2! = 10$ points of intersection since we have five constraints and two decision variables. $\binom{5}{2}$ is a combinatorial term that can be written in terms of factorials:

$$\binom{5}{2} = \frac{5!}{2!(5-2)!}$$

[5! is read "5 factorial" and is equal to the product $5(5-1) \ldots 2 \cdot 1$; the other factorials are computed in similar fashion.]

Pair 1:*	$40a + 30b = 480$	$a = 12/7$	$P = 118 \ 2/7$
	$24a + 32b = 480$	$b = 96/7$	
Pair 2:	$40a + 30b = 480$	$a = -8$	$P = \text{undefined}$
	$20a + 24b = 480$	$b = 80/3$	

Pair 3:	24 a + 32 b = 480	a = 60	P = undefined
	20 a + 24 b = 480	b = −30	
Pair 4:	40 a + 30 b = 480	a = 0	P = 128
	a = 0	b = 16	
Pair 5:*	40 a + 30 b = 480	a = 12	P = 60
	b = 0	b = 0	
Pair 6:*	24 a + 32 b = 480	a = 0	P = 120
	a = 0	b = 15	
Pair 7:	24 a + 32 b = 480	a = 20	P = 100
	b = 0	b = 0	
Pair 8:	20 a + 24 b = 480	a = 0	P = 160
	a = 0	b = 20	
Pair 9:	20 a + 24 b = 480	a = 24	P = 120
	b = 0	b = 0	
Pair 10:*	a = 0		P = 0
	b = 0		

The solution to each pair of equations must now be inserted into the original set of constraints to ensure that values of a and b are feasible (i.e., values of a and b do not violate the constraints). Recall that our constraints are:

$$40 \ a + 30 \ b \leqslant 480 \qquad \text{(Constraint 1)}$$

$$24 \ a + 32 \ b \leqslant 480 \qquad \text{(Constraint 2)}$$

$$20 \ a + 24 \ b \leqslant 480 \qquad \text{(Constraint 3)}$$

$$a \geqslant 0 \qquad \text{(Constraint 4)}$$

$$b \geqslant 0 \qquad \text{(Constraint 5)}$$

The solution to Pair 1 satisfies the three constraints (3, 4 and 5 above) not used to determine the point of intersection at a = 12/7 and b = 96/7. That is

$$20(12/7) + 24(96/7) \ < \ 480$$

$$12/7 \ > \ 0$$

$$96/7 \ > \ 0$$

Solutions to Pairs 2 and 3 are not permissible since a and b must be non-negative. After evaluating other points of intersection in the same manner, it is apparent that solutions to Pairs 1, 5, 6 and 10 satisfy all five constraints. Asterisks have been placed by each of these solutions.

These four points are termed *basic feasible* solutions to our linear programming problem and lie at the *vertices* of the feasible region formed by the constraints. The optimal solution will be located at one of these vertices (a single point). It is also true that all points lying on or within the feasible region are *feasible* solutions to the problem. If the objective function is coincident with (parallel to) one of the constraints there are an infinite number of solutions resulting in a profit of $P.

The above enumeration of intersection points illustrates that the optimal solution is the basic feasible solution that maximizes (or minimizes) the objective function. In our problem it is seen that the solution to Pair 6 yields the maximum profit of $120 at a = 0 and b = 15. Therefore, we would recommend that 15 units of Item b be produced each day.

In making this recommendation to management, we could carry the analysis one step further and calculate idle time in each department:

Department 1	$480 - 40(0) - 30(15) =$	30 minutes/day
Department 2	$480 - 24(0) - 32(15) =$	0 minutes/day
Department 3	$480 - 20(0) - 24(15) =$	120 minutes/day

We may now want to suggest that management consider the production of Item c (a new product line), which would require machine work mainly in departments 1 and 3. If this were possible we would formulate a new linear programming problem with three decision variables and six constraints. After reworking the problem we could then recommend whether it is profitable to introduce the new product line. This will be illustrated with the Simplex Method in the next chapter.

A GRAPHICAL

SOLUTION TO A

LINEAR

PROGRAMMING

PROBLEM

Let us consider the graphical solution to a problem having two decision variables. Suppose the Ajax Furniture Company buys its lumber from two sources, G and H, and grades the lumber into three different grades, A, B, and C.

The following matrix shows the expected percentage of lumber in each grade from each source. Board-feet requirements are also shown.

		Source		*Board-feet required*
		G	H	
	A	.15	.60	3000
Grade	B	.25	.30	2500
	C	.60	.10	2000

Company G charges $1 per board-foot and H charges $1.50 per board-foot. How much should be purchased from each supplier to satisfy requirements and minimize total cost?

This is a cost minimization problem that can be solved by using linear programming. We must first formulate our objective function and constraint equations. Let decision variable x_1 be the board-feet purchased from G, and x_2 the board-feet from H. The objective is to:

$$\text{Minimize } \$1.00\ x_1 + \$1.50\ x_2$$
$$\text{Subject to } .15x_1 + .60\ x_2 \geqslant 3000$$
$$.25x_1 + .30x_2 \geqslant 2500$$
$$.60x_1 + .10x_2 \geqslant 2000$$
$$x_1 \geqslant 0, x_2 \geqslant 0$$

The constraints and objective function are illustrated in Figure 4-5. Because we are minimizing costs, the optimal solution is the last point in the feasible region as the objective function moves *toward* the origin. This occurs at $x_1 = 5716$, and $x_2 = 3571$ with a minimum cost of $11,072. It can be seen that the slope of the objective function will change if the prices charged by G and H change. It can be seen from the slopes of the objective function and the constraint $0.25x_1 + 0.30x_2 \geqslant 2500$ that H will have to reduce his price to $1.20 to gain additional sales from our company.

Let us quickly consider the graphical solution to one more problem having two decision variables. Suppose the Pearl Air Cargo Company has just received a request to haul 800 tons of a perishable material from its home base to a town 1500 miles away. The company has two types of cargo planes, and their number and capacity are shown below.

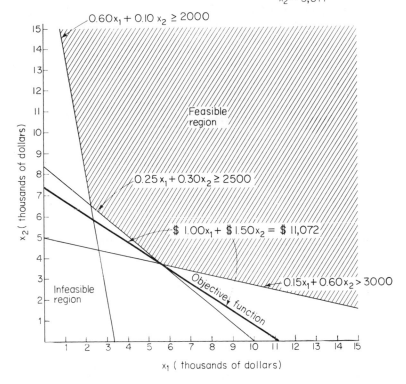

FIGURE 4-5   A Graphical Solution to the Lumber Problem

	Total Number	Capacity
Type I	5	150 tons
Type II	8	110 tons

The cost of operating Type I and Type II planes has been determined from historical data and is $0.06 and $0.04 per ton-mile, respectively. Maintenance procedures require that there be at least two Type II and one Type I plane on the ground at all times. Company policy requires that there always be in service at least one more Type II plane than Type I. The company requests your assistance in helping them decide how many planes of each type to use on this job so that operating costs are minimized. We will deal with one-way costs only because round trip costs are roughly twice the one-way costs.

This is a cost minimization problem that can be solved by using linear programming. To work this problem graphically, we must formulate our objective function and constraint equations. The decision variables are the number of Type I and Type II aircraft to be sent on this job. If we let x_1 equal the number of Type I aircraft and x_2 equal the number of Type II aircraft dispatched, the objective function is:

$$\text{Minimize } C = \frac{\$0.06}{\text{ton-mile}} (150 \text{ tons})(1500 \text{ miles}) +$$

$$\frac{\$0.04}{\text{ton-mile}} (110 \text{ tons})(1500 \text{ miles})$$

or

$$\text{Minimize } C = 13{,}500x_1 + 6600x_2$$

From the problem statement, we can write these constraints:

1. $150x_1 + 110x_2 \geqslant 800$ We must have the capability of hauling at least 800 tons of material.

2. $x_2 - x_1 \geqslant 1$ Company policy requires that there always be in service at least one more Type II plane than Type I.

3. $x_1 \leqslant 4$ No more than four Type I aircraft are available (one on ground at all times).

4. $x_2 \leqslant 6$ No more than six Type II aircraft are available (two on ground at all times).

5. $x_1 \geqslant 0$ }   Negative aircraft assignments are not possible.

6. $x_2 \geqslant 0$ }

We now plot these constraints and the objective function on an $x_1 - x_2$ graph as shown in Figure 4-6. Because we are minimizing costs, the optimal solution is the last point in the feasible region as the objective function moves toward the origin. This occurs at $x_1 = 1$ and $x_2 = 6$ and the minimum cost is $52,200.

With the graphical introduction to linear programming provided by this chapter, it is now possible to envision the process of determining optimal solutions. In the next chapter a systematic procedure is described for solving larger problems involving numerous decision variables and constraints.

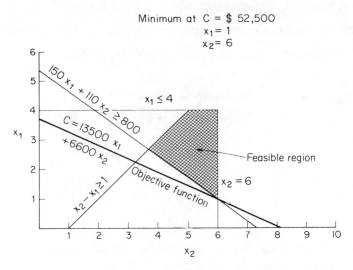

FIGURE 4-6   A Graphical Solution to the Air Cargo Problem

SUGGESTED ADDITIONAL READINGS

Smythe, W.R., and Johnson, L.A. *Introduction to Linear Programming with Applications.* Englewood Cliffs, N.J.: Prentice-Hall, Inc., 1966.

Draper, J.E., and Klingman, J.S. *Mathematical Analysis—Business and Economic Applications.* New York: Harper and Row, 1967.

Stockton, R. Stansbury. *Introduction to Linear Programming* (2nd Edition). Boston: Allyn and Bacon, 1963.

Llewellyn, R.W. *Linear Programming.* New York: Holt, Rinehart and Winston, 1964.

EXERCISES

1. Enumerate all basic feasible solutions for this problem:

$$\text{Maximize} \quad P = x_1 + 2x_2$$
$$\text{Subject to} \quad 2x_1 + 3x_2 \leqslant 12$$
$$5x_1 + 2x_2 \leqslant 15$$
$$x_1 \geqslant 0$$
$$x_2 \geqslant 0$$

(a) What is the optimal solution?

(b) Now find the optimal solution by using the graphical procedure.

2. Determine the feasible region for the following problem:

$$\text{Minimize} \quad C = 4x_1 + 6x_2$$
$$\text{Subject to} \quad 2x_1 + 5x_2 \geqslant 10$$
$$3x_1 + 2x_2 \geqslant 6$$
$$x_1 \geqslant 0$$
$$x_2 \geqslant 0$$

(a) Using graphical means, determine the optimal solution.

(b) If the objective function changes to $C = x_1 + 8x_2$, will there be a different optimal solution?

(c) The objective function, $C = x_1 + 8x_2$, represents a family of straight lines with what slope?

3. Suppose your boss asks you to maximize a process whose yield is given by $Z = 2x_1 + 3x_2$. Constraints on both controllable variables in the process are

$$-x_1 + 2x_2 \leqslant 16$$
$$-2x_1 + 2x_2 \leqslant 4$$
$$x_1 \geqslant 0$$
$$x_2 \geqslant 0$$

(a) Solve this problem graphically. What recommendation would you give your boss?

(b) How would you explain to him the difficulty you have encountered?

4. The Fleetwood Company mixes three different types of grass seed in their "Extemporal Blend," which sells for $0.80 per pound. Their less expensive blend, "Prolusion Blend," sells for $0.70 per pound and consists of two types of seed. The mixing formula for both blends is shown below.

Blends:	Seeds: Type A	Type B	Type C
Extemporal	½	¼	¼
Prolusion	0	¾	¼
Amount Available	400 lbs.	300 lbs.	500 lbs.

The cost per pound of Type A, Type B and Type C grass seed is $0.25, $0.21 and $0.18, respectively. Left over seed can be sold at this cost. The problem is to determine the optimal amount of each blend that should be prepared. Formulate this as a linear programming problem and solve it by using graphical methods.

5. A small company binds books and has two bindings available. Binding A is a high-quality product that results in a profit of $1.80 per book. Binding B is a lower quality product with a profit margin of $1.50 per book. If only the lower quality binding were available, the company could bind 500 books each day. When Binding A is requested, it requires 150 percent more time than does Binding B. However, because of material shortages only 350 books each day can be produced regardless of the type of binding. The high-quality binding requires a special glueing operation that has a maximum output of 250 books per day. Formulate this production problem as a linear programming problem to maximize profit and solve it graphically.

5

THE SIMPLEX METHOD OF LINEAR PROGRAMMING

INTRODUCTION

In Chapter 4 we saw how linear programming problems having two variables can be solved with graphical methods. Linear programming is an optimization tool applicable to special types of resource-allocation problems having these characteristics:

1. Decision variables are non-negative (the variables over which we have control—for example, quantities of four styles of transistor radios to manufacture—cannot be negative).

2. Limitations on available resources are expressed in terms of linear inequalities (for instance we may have on hand only a limited number of certain components required in manufacturing radios). These inequalities are termed constraints.

3. The criterion used to judge the "goodness" or merit of alternatives is a linear function of the decision variables (we may desire to maximize profits by specifying the optimal number of radios of each type to produce). A linear function involves only first powers of the variables and no cross-product terms.

We also observed that the set of points satisfying all the constraints and non-negativity conditions represented the feasible region for our solution. In addition, the optimum solution is located at a vertex (corner) of the feasible region or perhaps along an edge connecting two vertices. Vertices of the feasible region are also referred to as "extreme points." As we saw, the extreme point that maximizes the objective function can be found by moving the objective function as far as possible from the origin while still having at least one point in common with the feasible region. In a minimization problem we would move the objective function toward the origin until we had at least one point in common with the feasible region. Finally, we noted that a linear programming problem may not have a feasible solution when the constraint set is empty or when the objective function can be increased (decreased) indefinitely while satisfying the constraints.

Because a wide assortment of real-world problems can be formulated in terms of linear programming, this chapter is devoted to a description of the Simplex Method. This procedure greatly reduces the time required to solve large problems through an algebraic analog to graphical solution methods. We will first concentrate on fundamental principles underlying the Simplex Method and then turn our attention to the systematic use of a set of rules to follow in performing the required calculations. Throughout this chapter examples are employed to illustrate basic concepts of the Simplex Method, and an effort is made to minimize use of unnecessary symbolic notation.

ENUMERATION OF

BASIC FEASIBLE

SOLUTIONS

To exemplify some of the underlying principles of the Simplex Method, consider again the production problem of the previous chapter. Suppose that management decides to add a new product line (Item c) that requires 20 minutes of machining time in department 1 and 32 minutes of machining time in department 3. The constraints and corresponding feasible region for the problem would now be:

$$
\begin{array}{lllll}
(1) & 40a + 30b + 20c & \leqslant 480 & & \text{Department 1 Time} \\
(2) & 24a + 32b & \leqslant 480 & & \text{Department 2 Time} \\
(3) & 20a + 24b + 32c & \leqslant 480 & & \text{Department 3 Time} \\
(4) & a & \geqslant 0 & & \left.\begin{array}{l} \text{Non-negativity} \\ \text{constraints} \\ \text{on each product} \\ \text{line} \end{array}\right. \\
(5) & b & \geqslant 0 & & \\
(6) & c & \geqslant 0 & &
\end{array}
$$

The three-dimensional graph below shows the feasible region to be a polyhedron.

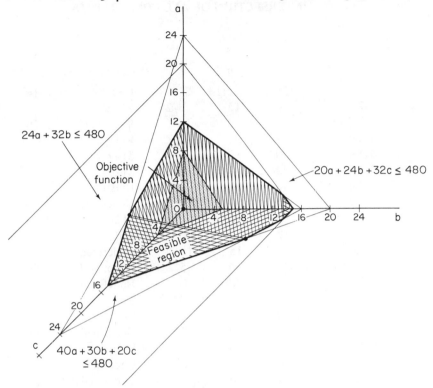

If the profit per unit of Item c were \$8, the objective function is $P = 5a + 8b + 8c$. This is the shaded plane in the graph whose location depends on the value of P and whose slope is determined by the coefficients of the objective function. Because we are attempting to maximize P, the plane would move away from the origin until the optimal solution is reached.

It is difficult to determine the point resulting in maximum profits from a three-dimensional graph. Apparently the optimum solution lies in the b-c plane and is equal to 12 units of Item b and 6 units of Item c. The profit would be \$144, which exceeds our maximum profit of \$120 with only two decision variables, a and b.

To determine the exact optimal solution, we would proceed to solve $\binom{6}{3}$ = 6!/3!3! = 20 sets of simultaneous equations and then select the feasible point with maximum P. Each set of three linear equations with three unknowns will have a single unique solution if the equations are linearly independent.

It is obvious by this point that enumerative search procedures would become extremely laborious for larger problems. However, because this same problem will be used in illustrating an intuitive development of the Simplex Method, solutions to these 20 sets of simultaneous equations and the corresponding values of P are listed in Table 5-1.

TABLE 5-1   EVALUATION OF THE
INTERSECTION OF ALL CONSTRAINTS

Constraint Equations Listed on P. 96	a	b	c	P = 5a + 8b + 8c	Constraint Violated
1,2,3	-36/13	222/13	51/13	Undefined	(4)
1,2,4	0	15	3/2	$132*	—
1,2,5	20	0	-16	Undefined	(6)
1,2,6	12/7	96/7	0	$118 2/7*	—
1,3,4	0	12	6	$144**	—
1,3,5	72/11	0	120/11	$120*	—
1,3,6	-8	80/3	0	Undefined	(4)
1,4,5	0	0	24	Undefined	(3)
1,4,6	0	16	0	$128*	—
1,5,6	12	0	0	$60*	—
2,3,4	0	15	15/4	Undefined	(1)
2,3,5	20	0	5/2	Undefined	(1)
2,3,6	60	-30	0	Undefined	(5)
2,4,5	0	0	0	$0*	—
2,4,6	0	15	0	$120*	—
2,5,6	20	0	0	Undefined	(1)
3,4,5	0	0	15	$120*	—
3,4,6	0	20	0	Undefined	(1,2)
3,5,6	24	0	0	Undefined	(1,2)
4,5,6	0	0	0	$0*	—

*Indicates basic feasible solution.
**Indicates the optimal solution.

Thus, we see that the optimal solution now is to produce 0 units of Item a, 12 units of Item b and 6 units of Item c at a profit of $144 per day. It should now be clear that evaluating the objective function at each corner of an n-dimensional polyhedron (n equals the number of decision variables) would be virtually impossible by manual methods for a large problem.

INTUITIVE
DEVELOPMENT OF
THE SIMPLEX
METHOD

A systematic procedure for searching among only those vertices of the feasible region that result in improved values of the objective function is needed. The

Simplex Method was developed around this principle and essentially consists of these steps:

1. Locate one vertex (corner) of the feasible region that we know will be a basic feasible solution. This is the current "best" solution. For example, setting all decision variables equal to zero will usually satisfy this requirement.

2. Examine other corners of the feasible region emanating from the starting point found in Step 1. If an improvement in the objective function is possible at one of these corners, it becomes our new "best" solution and we proceed to Step 3. If no improvement is possible, we have determined the optimal solution.

3. When the objective function can be bettered, choose the corner that permits the largest improvement to be made and call it the new "best" solution.

4. Repeat the above three steps until no further improvement can be made in the objective function.

The task now is to describe how each of the above steps is to be implemented in terms of a standard computational procedure. To begin we will convert all inequality constraints into equalities through the use of "slack" variables. This is an important first step because sets of equalities are much easier to work with than are sets of inequalities. In our manufacturing problem we would have one slack variable for each machining department as follows.

$$40a + 30b + 20c + d \qquad = 480$$
$$24a + 32b \qquad\quad + e \qquad = 480$$
$$20a + 24b + 32c \qquad + f \quad = 480$$

The slack variables, d, e and f, represent unused or excess time in departments 1, 2 and 3, respectively. For any feasible combination of a, b and c produced, each of the constraints above can be made an equality by adding the appropriate amount of excess time in a department. To illustrate, suppose that a = 5, b = 4 and c = 6 in department 1. Then d must equal 40 minutes for the equality to hold:

(40 min./unit)(5 units) + (30 min./unit)(4 units) +
(20 min./unit)(6 units) + 40 min. idle time = 480 minutes available

Our objective function becomes:

$$P = 5a + 8b + 8c + 0{\cdot}d + 0{\cdot}e + 0{\cdot}f$$

No profit results from idle time in each department so the profit coefficients of d, e and f are zeros. The above set of constraints could be written as:

$$40a + 30b + 20c + d + 0{\cdot}e + 0{\cdot}f \quad = \quad 480$$

$$24a + 32b + 0{\cdot}c + 0{\cdot}d + e + 0{\cdot}f \quad = \quad 480$$

$$20a + 24b + 32c + 0{\cdot}d + 0{\cdot}e + f \quad = \quad 480$$

If we choose as our initial solution a known feasible solution that in this case is the origin, then a = b = c = 0, d = e = f = 480 and P = 0. In matrix notation our first solution is

$$
\begin{bmatrix} 1 & 0 & 0 \\ 0 & 1 & 0 \\ 0 & 0 & 1 \end{bmatrix}
\begin{bmatrix} d \\ e \\ f \end{bmatrix}
=
\begin{bmatrix} 480 \\ 480 \\ 480 \end{bmatrix}
$$

In this particular problem we have three equations and six unknown quantities. As you recall from Chapter 3, there are n!/m!(n−m)! solutions to m equations and n unknowns (m < n) if the equations are linearly independent. That is, by setting (n−m) or more decision variables equal to zero and solving the resultant equations, the "basic solutions" to a linear programming problem can be determined. However, we saw in Chapter 4 and earlier in this chapter that not all basic solutions are feasible solutions. Frequently, one or more constraints are violated by some of the basic solutions. Those basic solutions satisfying all constraints are called "basic feasible solutions" to the linear programming problem and nonzero variables in these solutions are then termed "basic variables." Variables with a value of zero in a basic feasible solution are "nonbasic." In the example d, e and f are initially our basic variables and a, b and c are nonbasic variables.

Our next task is to select as a new "best" solution the vertex of the feasible region resulting in the greatest improvement of the objective function. This vertex must be one of the basic feasible solutions. If an improvement in the objective function is to occur, we need to make a, b or c some positive quantity since our initial solution has a profit of P = 5·0+8·0+8·0+0·480+0·480+0·480 = 0. To decide which nonbasic variable should become a basic variable, we must examine the consequences of bringing a nonbasic variable into our solution and *dropping* one of the basic variables from the present solution.

If we let Item a be a basic variable, the profit will be increased by $5 for each unit of Item a that is produced. Similarly, both Items b and c will result in unit profits of $8 if they become basic variables. Therefore, it makes sense to have decision variable b or c in the solution. Let us assume that decision variable c is brought into the solution. Which variable in our initial solution will be dropped? We must also be concerned with the *net* improvement in our objective function after c is made a basic variable and either d, e or f is dropped from the current solution. In the Simplex Method only one decision variable will be added and one other dropped at each iteration.

To answer the question regarding which decision variable will be dropped from the solution, let us reconsider the constraint set of our initial basic feasible solution.

$$
\left.
\begin{aligned}
40a + 30b + 20c + d & & = 480 \\
24a + 32b \quad\quad + e & & = 480 \\
20a + 24b + 32c \quad\quad + f & & = 480
\end{aligned}
\right\}
\quad
\begin{aligned}
& \text{Initial Solution:} \\
& d = e = f = 480 \\
& P = 0
\end{aligned}
$$

Because we have elected to make Item c a basic variable, we still do not know whether to drop d, e or f, but we do know that a = b = 0 (i.e., they will remain nonbasic variables). Thus, to maximize profits we want to produce as many units of Item c as possible without violating one (or more) of the constraints.

In department 1 we can produce $480/20 = 24$ units of Item c and in department 3 we can manufacture only $480/32 = 15$ units of Item c. (Note that Item c requires no time in department 2.) In this case no more than 15 units of Item c can be produced without violating the third constraint (f would become negative). Since the third constraint is the "tightest" limitation on the number of units of Item c that can be made, we are forced to bring Item c into our solution at a production level of 15 units. To solve for this value of c, we must make the transformation shown below:

$$
\begin{bmatrix} 20c \\ 0c \\ 32c \end{bmatrix} \longrightarrow \begin{bmatrix} 0c \\ 0c \\ 1c \end{bmatrix}
$$

In this case, we must have a "1" in the third constraint under decision variable c so that the maximum permissible value of c can readily be determined. This means that decision variable f, which currently has a "1" in this position, must leave the initial solution so that Item c can enter.

Because a and b are nonbasic variables, they are in our solution at a production level of 0 and we could write

$$
d = 480 - 20c
$$

$$e = 480$$

$$f = 480 - 32c$$

Suppose c is increased by one unit. Now all time in departments 1 and 3 will be accounted for if d = 460 and f = 448. The change in profit equals the extra profit from one more unit of Item c *less* the decrease in profit from losing 20 minutes of idle time in department 1 and 32 minutes of idle time in department 3. There is no profit associated with idle time, so when a single unit of Item c is brought into the solution the net increase in the objective function will be: +1 unit of item c ($8/unit) – 20 min. idle time in dept. 1 ($0/unit) – 32 min. idle time in dept. 3 ($0/unit) = + $8. The total gain by adding 15 units of Item c is $120.

To make Item c a basic variable, adjustments are necessary in the constraint set because available time in departments 1 and 3 is no longer equal to 480 minutes. For example, there will be 480 – 20(15) = 180 minutes of time still available in department 1 for the manufacture of Items a and/or b (if we decide to bring them into the solution). To accomplish the indicated transformation we will use the Gauss-Jordan elimination procedure. We first divide the third row by 32 to obtain:

$$5a/8 + 3b/4 + c + f/32 = 15$$

Now if the transformed third constraint above is multiplied by –20 and added to the first constraint listed on page 101, the result is

$$55\ a/2 + 15\ b + d - 5f/8 = 180$$

The second constraint is unchanged while the new third constraint is shown above. Because we will be referring back to this set of modified constraints, it is summarized as follows:

$$55a/2 + 15b \qquad + d \qquad\quad -5f/8 = 180$$
$$24a \quad + 32b \qquad\quad + e \qquad\qquad = 480$$
$$5a/8 + 3b/4 + c \qquad\qquad +f/32 = 15$$

This is the equation set for the *second* solution, and by inspection you can see that c = 15, d = 180 and e = 480. Therefore, 15 units of Item c would be produced, resulting in 180 minutes of idle time in department 1 and 480 minutes of idle time in department 2.

Basic variables have an associated column vector containing a "1" and all other elements "0." These vectors compose the "basis" of a linear programming problem and in our second solution we have these variables in the basis:

$$c = \begin{bmatrix} 0 \\ 0 \\ 1 \end{bmatrix}, \quad d = \begin{bmatrix} 1 \\ 0 \\ 0 \end{bmatrix} \quad \text{and} \quad e = \begin{bmatrix} 0 \\ 1 \\ 0 \end{bmatrix}$$

Notice that the basis is actually an m × m identity matrix consisting of only those vectors currently in our "best" solution.

We must next decide whether to bring another nonbasic variable into the basis so that the objective function might be further improved. Decision variable a or b would be a logical candidate, but it may also be possible to reintroduce f into the solution. To discover which variable improves the objective function the most, we again examine the net effect on profit from producing one unit of Items a, b and f.

If we increase Item a from 0 units to 1 unit produced, the per unit effect on profit can be easily determined. Refer to the tableau for the second solution on p. 102. If Item a changes from 0 to 1 unit, it is apparent that less time will be available in all three machining departments for the production of other goods. Let us analyze the effect of a reduction in time available in department 1 (i.e. the first constraint). Here you can see that available machining time would be ($180 - 55/2 = 152\ 1/2$) minutes after producing one unit of Item a. This, in turn, reduces decision variable d from 180 to 152 1/2. Because no profit is lost by reducing d by 55/2 minutes, the net gain in the objective function by making one unit of Item a is $5, the unit profit of Item a, less $0. The *total* effect on profit of increasing Item a by one unit would be (taking account of changes in all three constraints):

effect of nonbasic variable a	(first constraint) effect on basic variable d	(second constraint) effect on basic variable e
+1 ($5/unit)	− 55/2 ($0/unit)	− 24 ($0/unit)

(third constraint)
effect on basic variable c

−5/8 ($8/unit)

or $5 − $5 = 0. Thus, there is no advantage in bringing Item a into the solution.

We next examine in the same fashion changes in the objective function caused by unit increases in the remaining two nonbasic variables:

Item b: +1($8/unit)–15($0/unit)–32($0/unit)–3/4($8/unit)
  = +$2/unit

Item f: +1($0/unit)+5/8($0/unit)–0($0/unit)–1/32($8/unit)
  = –$0.25/unit

These same effects upon profit of each nonbasic variable could have been determined from the equation set for the second solution (page 102). If we use the constraints to solve for the value of basic variables in terms of all nonbasic variables and then substitute the resultant equations into the objective function, we obtain:

$$d = 180 - \frac{55}{2}a - 15b + \frac{5}{8}f,$$

$$e = 480 - 24a - 32b,$$

$$c = 15 - \frac{5}{8}a - \frac{3}{4}b - \frac{1}{32}f, \text{ and}$$

$$P = 5a + 8b + 8(15 - \frac{5}{8}a - \frac{3}{4}b - \frac{1}{32}f), \text{ or}$$

$$P = 120 + 0a + 2b - 0.25f$$

From the last equation, the effect on P from introducing one unit of a, b and f is immediately apparent and identical to our previous results. This clearly demonstrates that decision variable b should enter the solution to improve P the most. The amount of Item b that we ought to produce is limited by constraints on machining time and, as before, the tightest constraint will dictate the maximum possible value of b.

Before we formalize decision rules regarding entering and departing variables, let us complete the problem at hand. We now know that decision variable b should be introduced into the basis. To determine how much of Item b can be produced without violating one (or more) of the constraints we again divide available time in a department by the coefficient of b in each constraint equation (refer to page 102 for time available):

Department 1   180/15 = 12*

Department 2	$480/32 = 15$
Department 3	$15/0.75 = 20$

The departing basic variable is the one present in the constraint corresponding to the least valued finite, positive quotient above. Hence, the first constraint limits the amount of Item b that can be produced. Decision variable d is dropped from the basis to allow b to enter since d occupies the $[1\ 0\ 0]^T$ position in the basis that Item b must now have to become a basic variable.

The column vector for b in the last equation set was $[15\ 32\ \frac{3}{4}]^T$. We must transform this to a $[1\ 0\ 0]^T$ vector for Item b to enter the basis. After employing the Gauss-Jordan elimination procedure, the new equation set for our *third* solution would be:

$$\frac{11}{6}a + 1 \cdot b + 0 \cdot c + \frac{d}{15} + 0 \cdot e - \frac{1}{24} \quad f = 12$$

$$\frac{-208}{6}a + 0 \cdot b + 0 \cdot c - \frac{32}{15}d + 1 \cdot e + \frac{4}{3} \quad f = 96$$

$$\frac{-3}{4}a + 0 \cdot b + 1 \cdot c \quad \frac{1}{20}d + 0 \cdot e + \frac{1}{16} \quad f = 6$$

Now we can solve for values of basic variables as follows: $b = 12$, $e = 96$ and $c = 6$. The profit would be $P = 0(5) + 12(8) + 6(8) = \$144$.

To determine whether we should continue by introducing another nonbasic variable into the solution, let us investigate the change in profit resulting from unit increases in a, d and f (these are current nonbasic variables):

Item a: $+1(\$5) - 11(\$8)/6 + 208(\$0)/6 + 3(\$8)/4 \quad = -\$3\frac{2}{3}$

Item d: $+1(\$0) - 1(\$8)/15 + 32(\$0)/15 + 1(\$8)/20 \quad = -\$2/15$

Item f: $+1(\$0) + 1(\$8)/24 - 4(\$0)/3 - 1(\$8)/16 \quad = -\$1/6$

Therefore, no increase in the objective function is possible by bringing a nonbasic variable into the basis and we have discovered the optimal solution. This solution, i.e. $b = 12$, $c = 6$, is identical to our previous one determined by explicitly evaluating each corner of the feasible region.

FORMALIZATION

OF THE SIMPLEX

METHOD

Now that we have seen an intuitive procedure for solving linear programming problems, we will formalize a sequence of computational steps that makes up the Simplex Method. The Simplex Method is an iterative technique that starts with a known basic feasible solution, and if this solution is not optimal it proceeds to optimize the objective function through the introduction of new decision variables. Because each successive solution (iteration) improves upon the current one, it is not possible to consider the same solution twice and the procedure terminates in a finite number of iterations. Thus, the Simplex Method is often called an "algorithm" since it embodies a sequence of instructions, or rules, arranged in a logical order that leads directly to a correct answer.

In general terms, we desire to maximize (minimize) a linear objective function subject to linear constraints on the decision variables. Mathematically, the problem statement is:

$$\text{Maximize} \quad Z = c_1 x_1 + c_2 x_2 + \cdots + c_n x_n$$

$$\text{Subject to} \quad a_{11} x_1 + a_{12} x_2 + \cdots + a_{1n} x_n \leq b_1$$

$$a_{21} x_1 + a_{22} x_2 + \cdots + a_{2n} x_n \leq b_2$$

$$\vdots \qquad \vdots \qquad \vdots \qquad \vdots$$

$$a_{m1} x_1 + a_{m2} x_2 + \cdots + a_{mn} x_n \leq b_m$$

$$x_j \geq 0$$

where c_j = profit per unit of $x_j (1 \leq j \leq n)$

x_j = decision variables or "activities"

a_{ij} = amount of resource i required by activity j $(1 \leq i \leq m)$

b_i = total amount of resource i available.

The following steps in the Simplex Method apply to maximization problems, but with a few minor changes (to be discussed later) minimization problems can also be solved.

**step 1—add slack
variables**

Convert all linear inequalities (constraints) to equalities through the addition of non-negative slack variables to the smaller side of the inequalities. If there are m restrictions on resource availability, we would add m slack variables to the problem. Then the problem would have n + m decision variables and m constraints. To illustrate each step of the Simplex Method, consider this linear programming problem.

$$\text{Maximize} \quad 4x_1 + 3x_2$$

$$\text{Subject to} \quad x_1 + 2x_2/3 \leqslant 6000$$

$$x_1 \qquad\qquad \leqslant 4000$$

$$x_2 \leqslant 6000$$

$$x_1, x_2 \qquad \geqslant 0$$

The first step would be to add a slack variable for each constraint to the smaller side of the inequality. Considering the first constraint, we would add a non-negative slack variable x_3 to the left-hand side of the inequality to form an equality: $x_1 + (2/3) x_2 + x_3 = 6000$. If the inequality had been reversed, i.e. $x_1 + (2/3) x_2 \geqslant 6000$, we would have written $x_1 + (2/3) x_2 = 6000 + x_3$, or $x_1 + (2/3) x_2 - x_3 = 6000$. This latter situation offers some difficulties that are treated in a later section. Because slack variables represent unused resources, their coefficients in the objective function are zero. After adding slack variables, we have the following:

$$\text{Maximize P} = \quad 4x_1 + 3x_2 + 0 \cdot x_3 + 0 \cdot x_4 + 0 \cdot x_5$$

$$\text{Subject to} \quad x_1 + (2/3)x_2 + x_3 \qquad\qquad = 6000$$

$$x_1 \qquad\qquad + x_4 \quad = 4000$$

$$x_2 \qquad\qquad + x_5 - 6000$$

$$x_j \geqslant 0$$

Because it is convenient to solve these equations by using matrix algebra, we could write the problem as follows:

$$\begin{bmatrix} 1 & 2/3 & 1 & 0 & 0 \\ 1 & 0 & 0 & 1 & 0 \\ 0 & 1 & 0 & 0 & 1 \end{bmatrix} \begin{bmatrix} x_1 \\ x_2 \\ x_3 \\ x_4 \\ x_5 \end{bmatrix} = \begin{bmatrix} 6000 \\ 4000 \\ 6000 \end{bmatrix}$$

step 2—construct tableau

After putting the constraints in matrix form, construct a "tableau" that summarizes all information relevant to the problem. The tableau shows (a) which decision variable is associated with each column in the matrix, (b) available resources corresponding to each constraint, (c) profit (or cost) coefficients of all decision variables and (d) the variables in the basis. The initial tableau (still incomplete at this point) for our illustrative problem would be:

c_i BASIS	V_1	V_2	V_3	V_4	V_5	b_i	b_i/a_{ij}
	1	$\frac{2}{3}$	1	0	0	6000	
	1	0	0	1	0	4000	
	0	1	0	0	1	6000	
c_j ΔOF_j	4	3	0	0	0	—	

This is the format of linear programming tableaus that we will be using with the Simplex Method. At this point the tableau contains the three constraints (with slack variables) and the objective function. For instance, the first constraint is seen to be $x_1 + 2x_2/3 + x_3 = 6000$ while the objective function is $P = 4x_1 + 3x_2 + 0x_3 + 0x_4 + 0x_5$. The column vector V_j indicates which decision variable is associated with each column of coefficients in the matrix and c_j is the coefficient of the jth variable in the objective function. Other terms are defined in subsequent steps.

step 3—an initial feasible solution

The iterative process of determining an optimal solution to a linear programming problem is initiated by selecting a known basic feasible solution. To satisfy all constraints with certainty, it is customary to use the origin as our first basic feasible solution so long as available resources are non-negative. At the origin we have all nonslack variables set to zero, which means each slack variable is equal to resources available in its respective constraint. In other words, this is the "do nothing" alternative that results in an objective function value of zero. If another basic feasible solution is known we could use it as a starting point and possibly save ourselves many calculations.

After recording our initial basic feasible solution in the tableau, we have:

Initial Solution:

c_i	BASIS	V_1	V_2	V_3	V_4	V_5	b_i	b_i/a_{ij}
0	x_3	1	$\frac{2}{3}$	1	0	0	6000	
0	x_4	1	0	0	1	0	4000	
0	x_5	0	1	0	0	1	6000	
	c_j ΔOF_j	4	3	0	0	0		

In using the origin as a first solution, you can see that $x_1 = 0, x_2 = 0, x_3 = 6000, x_4 = 4000$ and $x_5 = 6000$ in the example problem. The basis consists of $[V_3 \ V_4 \ V_5]$, which is an identity matrix, and x_3, x_4 and x_5, which are the basic variables. Thus, we determine the solution in each tableau from the nonzero variables whose column vectors consist of a 1 and all other entries are 0.

Notice that in the tableau we have more variables than constraint equations. As you recall from Chapter 3, this situation usually results in infinitely many solutions. In fact the feasible region of linear programming problems consists of infinitely many solutions, and the Simplex Method is an efficient procedure for determining the point (or perhaps points) in the feasible region that optimizes the objective function subject to a set of m linearly independent equations. This point, of course, will be one of the basic feasible solutions to the problem. Furthermore, there are more variables than constraints so we can have no more than m=3 nonzero variables in the optimal solution for the above problem. This means that we can expect to see unused resources in at least one constraint (because two nonslack activity variables and three constraints on resources permit one slack variable to enter this optimal solution).

**step 4—select
entering variable**

We must now check whether the solution for this (i.e. the initial) iteration is optimal. This is accomplished by calculating a "Simplex criterion" for each *nonbasic* variable. The criterion relates the net change in our objective function per unit of a nonbasic variable hypothetically introduced into the basis. The variable causing the greatest improvement in the objective function is selected to be the incoming, or entering, variable and (for the moment) we ignore all other nonbasic variables. Only one variable will be dropped from the basis at each iteration.

The Simplex criterion for *nonbasic variable* x_j is determined with this expression:

$$\Delta OF_j = c_j - \sum_{i=1}^{m} a_{ij} \, c_i$$

where ΔOF_j = value of the Simplex criterion for nonbasic variable j (represents net change in the objective function)

c_j = coefficient of nonbasic variable x_j in the objective function

c_i = objective function coefficients of basic variables, obtained from c_i column of Simplex tableau (c_1 is the first element of the c_i vector, c_2 the second element, etc.)

a_{ij} = amount of resource i (from the requirements matrix) consumed by variable j

To illustrate the calculation of ΔOF_j, consider nonbasic variable x_1. Here we have ΔOF_1 = 4-1(0)-1(0)-0(0) = 4 since a_{11} = 1, a_{21} = 1, a_{31} = 0. Also $c_{i=1}$, $c_{i=2}$ and $c_{i=3}$ all equal zero because there is no profit associated with unused resources. Similarly, ΔOF_2 = 3-2/3 (0) -0(0)-1(0) = 3. You can see ΔOF_1 > ΔOF_2, which means that x_1 should be brought into the basis. If all criteria had been negative no further improvement in the objective function would have been possible and a unique optimal solution would have been determined in Step 3.

Occasionally, a Simplex criterion for a nonbasic variable will have a value of zero. In this case it is possible to substitute the nonbasic variable for one of the basic variables with no change in the objective function. Hence the existence of a zero-valued Simplex criterion for a nonbasic variable means there are *multiple solutions* to the problem. In geometric terms, multiple solutions occur when the objective function is coincident with (parallel to) an active constraint so that infinitely many points are optimum solutions to the problem.

Finally, another difficulty can be encountered when two (or more) nonbasic variables have identical values of ΔOF_j. It is suggested that either variable can be chosen arbitrarily to enter the basis when ties are observed.

When one (or more) of the Simplex criteria is positive, an improvement in the objective function can be made by introducing a new variable into the basis. To make the greatest improvement, the entering variable should be the nonbasic variable having the largest ΔOF_j. In addition this variable must have at least one a_{ij} greater than zero (the reason will be apparent in Step 6).

To bring a new variable into the solution, we must transform its column vector (V_j) into a vector having all zero coefficients except for a single coefficient of 1. We shall repeatedly refer to the column vector of the incoming variable as the "pivotal column." Because we can transform the pivotal column to its desired form by using the Gauss-Jordan elimination procedure, we need to know which of the m elements in this column is to become a 1 (all other elements will become zeros). This element is determined in Step 5.

step 5—select
departing variable

Based on our knowledge of the entering variable from Step 4, we also must choose a departing variable. The row containing our departing variable is called the "pivotal row," and its selection is made so we are certain that values of all basic variables in the solution are non-negative. As described in the previous section, the row containing the departing variable is the one that limits the amount by which the entering variable can be increased. If the entering variable is increased by more than this restraining amount, one or more of the current basic variables will become negative and the solution will be infeasible.

To explain this concept in terms of our example problem, let us consider the nonzero variables in the last tableau (x_1 is no longer set at zero since it was chosen to enter the basis in Step 4):

$$x_1 + x_3 \quad\quad = 6000$$

$$x_1 \quad\quad + x_4 \quad = 4000$$

$$x_5 = 6000$$

It should be apparent that in the first constraint x_1 can be increased to 6000 without making x_3 negative. From the second constraint you will see that x_1 cannot be greater than 4000 without causing x_4 to become negative. Therefore, the second constraint is the "tightest" and dictates the value of x_1 in the new solution.

The pivotal row could have been determined by examining the ratios of constant terms in the requirements vector, $[b_1 \quad b_2 \quad \cdots \quad b_m]^T$, to the corresponding a_{ij} in the column vector of the entering variable. A decision is made to remove the vector whose b_i/a_{ij} is the *smallest positive value*, where j indexes the column of the entering variable. (We divide by a_{ij} because a 1 must appear in the column vector of the new basic variable and all other entries in this vector must be zeros.)

An updated version of the initial tableau appears below (based on information from Steps 4,5). The tableau shows that x_1 (associated with V_1) enters the basis and x_4 (corresponding to V_4 which had been in the basis) is leaving. The intersection of the pivotal column and pivotal row is normally called the "pivotal element." In our problem the coefficient in this position is $a_{21} = 1$. It is next necessary to transform the coefficient matrix so that the pivotal element equals 1 and all other elements in the pivotal column are zero.

c_i	BASIS	V_1	V_2	V_3	V_4	V_5	b_i	b_i/a_{ij}
0	x_3	1	$\frac{2}{3}$	1	0	0	6000	6000
0	x_4	(1)	0	0	1	0	4000	4000 $\longrightarrow$ departing variable, x_4
0	x_5	0	1	0	0	1	6000	∞
	c_j	4	3	0	0	0		—
	ΔOF_j	4	3	0	0	0		$\boxed{0}$

value of objective function in initial tableau = CB = $\Sigma c_i b_i$

pivotal element

entering variable, x_1

Before moving to Step 6, two comments should be made about interpreting the value of b_i/a_{ij}. Because the b_i are to be non-negative by definition, the algebraic sign of a_{ij} is important. If $a_{ij} = 0$ division is undefined, and if $a_{ij} < 0$ it would be possible to increase x_j indefinitely so that the objective function would have a value of $+ \infty$ ($-\infty$) in a maximization (minimization) problem. Therefore, coefficients in the column vector of our incoming variable must consist of at least one $a_{ij} > 0$. Otherwise, the real-world significance of the linear programming problem would be questionable.

The second comment concerns the possibility of having two (or more) rows with the same positive value of b_i/a_{ij}. An arbitrary choice of the pivotal row can result in a condition known as "cycling" in which there is an endless looping through a set of basic feasible solutions with no convergence toward an optimal solution. It is then possible that a solution will not be determined by using the Simplex Method. To avoid this difficulty, it is recommended that ties in b_i/a_{ij} quotients be broken by selecting for a pivotal row the one having the *largest* a_{ij} in its pivotal column.

step 6—apply Gauss-Jordan Procedure

To obtain a column vector with a 1 as the pivotal element and zeros elsewhere, we shall use the Gauss-Jordan elimination procedure discussed in Chapter 3. After completing the transformation, column vectors of the basic variables will constitute an m-dimensional identity matrix. In addition it is necessary to update the entries in the "c_i" and "BASIS" columns of the tableau. The transformed coefficient matrix for the example problem is this. (Notice that x_1 has replaced x_4 in the BASIS column. The coefficient of x_4 in the objective function (4) is entered in the c_i column.)

Second Iteration (Incomplete):

c_i	BASIS	V_1	V_2	V_3	V_4	V_5	b_i	b_i/a_{ij}
0	x_3	0	$\frac{2}{3}$	1	-1	0	2000	
4	x_1	1	0	0	1	0	4000	
0	x_5	0	1	0	0	1	6000	
	c_j	4	3	0	0	0	—	
	ΔOF_j						16,000	$= \Sigma c_i b_i$

Because there already was a 1 as the pivotal element (i.e., $a_{21} = 1$), it was necessary only to replace the first row of the old tableau by (Row 2 of *new* tableau) (-1) + (Row 1 of old tableau) to determine the first row of the *new* tableau. Otherwise, we would have first divided every element in the pivotal row by the value of the pivotal element. Thus the solution after the second iteration is $x_1 = 4000$, $x_3 = 2000$ and $x_5 = 6000$. The objective function has a value of P $= \Sigma c_i b_i = 4000(4) = 16,000$.

step 7—determine an optimal solution

The Simplex Method has now been formalized as a sequence of computational steps. To determine the optimal solution to a linear programming problem, we must repeat Steps 4-6 as necessary. For our problem, we would check the Simplex criteria as outlined in Step 4. If one or more criteria are positive we would move on to Step 5, and so forth. If all criteria had been negative our optimal solution could be read from the above tableau. To complete the problem at hand, the following tableaus are offered.

Second Iteration (Complete):

c_i	BASIS	V_1	V_2	V_3	V_4	V_5	b_i	b_i/a_{ij}	
0	x_3	0	②⁄₃	1	-1	0	2000	3000	→ departing variable
4	x_1	1	0	0	1	0	4000	∞	
0	x_5	0	1	0	0	1	6000	6000	
	c_j	4	3	0	0	0	—		value of objective
	ΔOF_j	0	3	0	-4	0	16,000		function in second tableau

entering
variable

Third Iteration:

c_i	BASIS	V_1	V_2	V_3	V_4	V_5	b_i	b_i/a_{ij}	
3	x_2	0	1	$\frac{3}{2}$	$-\frac{3}{2}$	0	3000	negative	
4	x_1	1	0	0	1	0	4000	4000	
0	x_5	0	0	$-\frac{3}{2}$	③⁄₂	1	3000	2000	→ departing variable
	c_j	4	3	0	0	0	—		value of objective
	ΔOF_j	0	0	$-\frac{9}{2}$	$\frac{1}{2}$	0	25,000		function in third tableau

entering
variable

Note: Here the first row of the second tableau was multiplied by 3/2. The second row of the second tableau is unchanged because $a'_{22} = 0$ (the prime designates the transformed a_{22} of the original matrix). Our third row above was obtained as follows—first row of third tableau (-1) + third row of second tableau. x_4 re-enters the basis and x_5, with an objective function coefficient of 0, is chosen to leave the basis.

Fourth Iteration: Optimal Solution

c_i	BASIS	V_1	V_2	V_3	V_4	V_5	b_i	b_i/a_{ij}
3	x_2	0	1	0	0	1	6000	
4	x_1	1	0	1	0	$-\frac{2}{3}$	2000	
0	x_4	0	0	-1	1	$\frac{2}{3}$	2000	
	c_j	4	3	0	0	0	—	value of objective
	ΔOF_j	0	0	-4	0	$-\frac{1}{3}$	26,000	function in fourth tableau

After performing the Gauss-Jordan elimination procedure and checking the Simplex criteria, we find that no improvement can be made in the objective function by adding a nonbasic variable to the solution. Thus, the optimal solution can be read after the fourth iteration to be $x_1 = 2000$, $x_2 = 6000$ and $x_4 = 2000$. Because x_4 is the slack variable for the second constraint and $x_3 = x_5 = 0$ (they are nonbasic variables in the above tableau), we have

$$\text{MAX P} = 4(2000) + 3(6000) = 26,000$$

Resources Utilized
$$\begin{cases} \text{Constraint 1:} \quad 2000 + (\frac{2}{3})\,(6000) \quad = \quad 6000 \\ \text{Constraint 2:} \quad 2000 + 2000 \qquad\quad = \quad 4000 \\ \text{Constraint 3:} \qquad\qquad\quad 6000 \qquad\quad = \quad 6000 \end{cases}$$

One last comment regarding this example should be made. In the original problem statement, there were two decision variables and three constraints (page 107). Consequently, the optimal solution could contain *no more than* m=3 nonzero basic variables. Our final solution, therefore, would include at most two of the initial decision variables (x_1 and x_2, x_2 or x_1 or neither) and it should not be surprising to observe $x_4 = 2000$ in the optimal solution. When a slack variable appears as a basic variable in the optimum solution, it simply means that the corresponding constraint does not restrain the solution and resources available can be reduced without changing the solution.

MINIMIZATION

VERSUS

MAXIMIZATION

PROBLEMS

The Simplex Method as described in the previous section is tailored to maximization problems. However, in solving *minimization* problems the same procedure can be followed subject to minor modifications. The first possibility is to multiply the objective function to be minimized by a -1 and then proceed to follow the Simplex Method as presented above. This can be done since maximizing the negative of a function is identical to minimizing the positive function. A second possibility is to choose the entering variable as that one with the most *negative* ΔOF_j in the above procedure. In this case an optimal solution would be determined from the tableau having all positive and/or zero Simplex criteria.

Still one difficulty remains in most minimization problems. That is the fact that the constraints are of the "greater than or equal" ($\geqslant$) type. Because we desire to minimize the objective function, constraints are often placed on the lower bound of an activity (decision variable) or set of activities. For example, production of a particular commodity must exceed R units per week to meet our sales commitments.

To deal with "$\geqslant$" or "=" types of constraints, it is customary that we introduce *artificial variables* into the linear programming problem. In essence artificial variables are needed to provide an initial basic feasible solution for the Simplex Method. Since these variables represent no tangible quantity, we set out to eliminate them as quickly as possible from the basis by giving each a highly undesirable coefficient in the objective function.

Before illustrating the use of artificial variables in an example problem, it must be noted that they can be employed whenever there is not an obvious (or otherwise known) basic feasible solution to use in commencing the Simplex Method. Thus, we may find it necessary to introduce artificial variables in maximization and minimization problems.

Let us consider the following linear programming problem:

$$\text{Minimize} \quad C = 2x_1 + 3x_2$$

$$\text{Subject to} \quad x_1 + x_2 = 4$$

$$6x_1 + 2x_2 \geqslant 8$$

$$x_1 + 5x_2 \geqslant 8$$

$$x_1, x_2 \geqslant 0$$

This minimization problem has both "$\geqslant$" and "$=$" constraints. After we add slack variables to the smaller side of the inequalities to get equalities, we have these constraints:

$$\left. \begin{array}{l} x_1 + x_2 = 4 \\ 6x_1 + 2x_2 = 8 + x_3 \\ x_1 + 5x_2 = 8 + x_4 \end{array} \right\} \quad \text{or} \quad \begin{array}{l} x_1 + x_2 \qquad\qquad = 4 \\ 6x_1 + 2x_2 - x_3 \qquad = 8 \\ x_1 + 5x_2 \qquad - x_4 = 8 \end{array}$$

In this case you can see that x_3 and x_4 would more appropriately be called "surplus" variables since they indicate how much of resource i must be subtracted from the left-hand side of the equation for the equality to hold. (Recall that *slack* variables relate resources unused on the left-hand side of the equation.)

It is virtually impossible to guess at a basic feasible solution to this problem, and we cannot utilize x_3 and x_4 as initial basic variables because then $x_3 = -8$ and $x_4 = -8$. Negative amounts of resources available (i.e. negative-valued decision variables) are not permissible. We need three positive-valued basic variables to initiate the Simplex procedure, so artificial variables are introduced for this purpose. Moreover, we must make these variables highly injurious in view of the objective function that we desire to minimize. This can be accomplished by assigning a very large value to each artificial variable in the objective function. If we let this value equal M, the problem becomes:

$$\text{Minimize C} = 2x_1 + 3x_2 - 0 \cdot x_3 - 0 \cdot x_4 + M(x_5 + x_6 + x_7)$$

$$\text{Subject to} \quad \begin{array}{l} x_1 + x_2 \qquad\qquad\quad + x_5 \qquad\qquad = 4 \\ 6x_1 + 2x_2 - x_3 \qquad\qquad + x_6 \qquad = 8 \\ x_1 + 5x_2 \qquad - x_4 \qquad\qquad + x_7 = 8 \\ x_j \geqslant 0 \ (1 \leqslant j \leqslant 7) \end{array}$$

(In maximization problems, we would use $-M$ for the coefficient of an artificial variable.) Now it is possible to use the Simplex Method by letting x_5, x_6 and x_7 be the basic variables, and our first tableau is shown below.

Initial Solution: First Tableau

c_i	BASIS	V_1	V_2	V_3	V_4	V_5	V_6	V_7	b_i	b_i/a_{ij}	
M	x_5	1	1	0	0	1	0	0	4	4	
M	x_6	⑥	2	-1	0	0	1	0	8	$4/3$	→ departing variable
M	x_7	1	5	0	-1	0	0	1	8	8	
	c_j	2	3	0	0	M	M	M	—		
	ΔOF_j	2	3	M	M	0	0	0	20M		
		-8M	-8M								

↑
entering variable

In a minimization problem such as this one, we could have set out to maximize $C = -2x_1 - 3x_2 - M(x_5 + x_6 + x_7)$. Instead, we have elected to reverse our interpretation of ΔOF_j so that now our incoming vector will have the *most negative* ΔOF_j. When all ΔOF_j are positive or zero, we have determined the optimal solution.

Our most negative ΔOF_j is associated with V_1 and hence x_1 is the entering variable. We must remove x_5, x_6 and x_7 from the solution by introducing legitimate decision variables. In the first tableau we removed x_6, although we could have started with any other artificial variable. After performing the Gauss-Jordan procedure in the next three tableaus, you will notice that all of the

Second Tableau:

c_i	BASIS	V_1	V_2	V_3	V_4	V_5	V_6	V_7	b_i	b_i/a_{ij}	
M	x_5	0	$2/3$	$1/6$	0	1		0	$8/3$	4	
2	x_1	1	$1/3$	$-1/6$	0	0		0	$4/3$	4	
M	x_7	0	⑭⑶ $14/3$	$1/6$	-1	0		1	$20 2/3$	$10/7$	→
	c_j	2	3	0	0	M		M	—		
	ΔOF_j	0	$7/3$	$1/3$	M	0		0	$8/3$		
			$-16/3 M$	$\dfrac{-M}{3}$					$+\dfrac{28}{3} M$		

↑

artificial variables are permanently eliminated from the problem (x_6 is dropped from the second tableau, x_7 is dropped from the third tableau and x_5 is dropped from the fourth tableau):

Third Tableau:

c_i	BASIS	V_1	V_2	V_3	V_4	V_5	V_6	V_7	b_i	b_i/a_{ij}
M	x_5	0	0	$\left(\frac{1}{7}\right)$	$\frac{1}{7}$	1			$1\,\frac{2}{7}$	12 →
2	x_1	1	0	$-\frac{13}{84}$	$\frac{1}{14}$	0			$\frac{6}{7}$	Negative
3	x_2	0	1	$\frac{1}{28}$	$-\frac{3}{14}$	0			$1\,\frac{0}{7}$	40
	c_j	2	3	0	0	M			—	
	ΔOF_j	0	0	$\frac{1}{4}$ $-\frac{M}{7}$	$\frac{1}{2}$ $-\frac{M}{7}$	0			$\boxed{6 + \frac{12}{7}M}$	

Fourth Tableau:

c_i	BASIS	V_1	V_2	V_3	V_4	V_5	V_6	V_7	b_i	b_i/a_{ij}
0	x_3	0	0	1	1				12	
2	x_1	1	0	0	$\frac{1}{4}$				3	
3	x_2	0	1	0	$-\frac{1}{4}$				1	
	c_j	2	3	0	0				—	
	ΔOF_j	0	0	0	$\frac{1}{4}$				$\boxed{9}$	

In the fourth tableau all ΔOF_j are positive or zero, so our optimal solution is $x_1 = 3$, $x_2 = 1$, $x_3 = 12$ and $C = 3(2) + 3(1) = 9$. If, after dropping all artificial variables, Simplex criteria had been negative for one or more nonbasic variables we would have proceeded to calculate additional tableaus until the optimal solution had been determined.

DEALING WITH

COMPLICATIONS

In using the Simplex Method, various complications can arise. With constraints of the ">" and "=" type, we discussed the need for artificial variables and we saw how to deal with them. For example, if a constraint were $3x_1 + 8x_2 \geq 16$, we could have multiplied it by a -1 to obtain $-3x_1 - 8x_2 \leq -16$. Here you see that the "$\leq$" inequality is suitable in a maximization problem (for which our Simplex Method is designed), but the -16 could easily result in a negative valued decision variable. Consequently, an artificial variable (x_4) in addition to the surplus variable (x_3) would be introduced as follows:

$$3x_1 + 8x_2 - x_3 + x_4 = 16$$

We also noted that when nonbasic variables have the same positive ΔOF_j (in maximization problems), an arbitrary selection of the incoming variable can be made. If b_i/a_{ij} for two or more outgoing variables are tied, we select the variable associated with the largest a_{ij} in the pivotal column to leave the basis. In this way we attempt to avoid the remote possibility of cycling caused by *degeneracy* in our linear programming problem. (A degenerate solution is a basic feasible solution that has fewer than m nonzero basis variables in it. That is, one or more basic variables have a value of 0 in a solution. It is possible for a degenerate solution to be our optimal solution.)

The existence of multiple solutions to a linear programming problem was observed when the ΔOF_j of one or more nonbasic variables equals 0. This indicates that a nonbasic variable can be substituted for one of the basic variables in an optimal solution and the value of the objective function will not change. Thus, we discover that there can be a large number of optimal solutions to the problem.

Another complication can arise when all a_{ij} in the pivotal column are zero or negative. Because there would be no positive b_i/a_{ij} (assuming $b_i \geq 0$), we could not select the departing variable. In this situation there is no upper bound on the requirements vector (b_i) and, therefore, also no upper value for the objective function in a maximization problem. We should check the constraint set for errors in the direction of our inequalities (e.g. "$\leq$" instead of "$\geq$") when unbounded solutions are discovered.

To check for errors in constraints, we could compute the determinate of a nontrivial "basic feasible" solution (i.e., a solution obtained by means other than setting all slack or artificial variables to zero). For small problems this is not difficult but for large ones it would be. We might suspect an incorrectly formulated problem when it is impossible to remove an artificial variable from our tableaus. Such illogical events should cause suspicion regarding the formulation of the problem.

One last potential difficulty should be mentioned. Even though most decision variables are restricted to non-negative quantities in the Simplex Method, it is conceivable to have a problem in which one or more variables are unconstrained in sign. These situations can be transformed into equivalent

Example 1 **121**

problems involving only non-negative variables, which we can solve through conventional means. To illustrate, let us express a variable unconstrained in sign (x_1) as the difference between two non-negative variables:

$$x_1 = S_1 - t_1, S_1 \geqslant 0 \text{ and } t_1 \geqslant 0$$

We will then let $x_1 = S_1$ and $t_1 = 0$ when $x_1 \geqslant 0$. Or we will let $t_1 = -x_1, S_1 = 0$ when $x_1 \leqslant 0$. If we want to maximize $P = 2x_1 + 3x_2$ subject to $5x_1 + 3x_2 \leqslant 12$; $x_1 \leqslant 4; x_2 \leqslant 5; x_2 \geqslant 0$ when x_1 is unconstrained in sign, we could write:

$$
\begin{aligned}
\text{Maximize } P \quad &= \quad 2(S_1 - t_1) + 3x_2 \\
\text{Subject to} \quad &\quad 5(S_1 - t_1) + 3x_2 \leqslant 12 \\
&\quad S_1 - t_1 \leqslant 4 \\
&\quad x_2 \leqslant 5 \\
&\quad S_1, t_1, x_2 \geqslant 0
\end{aligned}
$$

Because the Simplex Method only examines basic feasible solutions, only S_1 or t_1 (or neither) would be included in a solution to the problem.

To illustrate further the formulation and solution of linear programming problems with the Simplex Method, two example problems are worked and explained in detail.

EXAMPLE 1

Suppose an agricultural extension agent is working with a local farmers' co-operative. The co-operative has a 100-acre farm. They can sell tomatoes at 19¢ a pound, lettuce at 10¢ a head and squash at 25¢ a pound. The average yield per acre is 2000 pounds of tomatoes, 3000 heads of lettuce and 1000 pounds of squash. Labor required for sowing, cultivating and harvesting each acre is five man-days for tomatoes and squash, and six man-days for lettuce. Within the co-operative, 400 man-days of labor are available and each laborer who is a member of the cooperative is paid $20 per man-day. Fertilizer must also be used and it costs 10¢ per pound. Requirements for fertilizer are these: 100 pounds per acre of tomatoes and lettuce, and 50 pounds per acre of squash. Based on this information, we desire to find the best way to use land and manpower to maximize profits. The following points are treated in our solution.

(a) Set this up as a linear programming problem.

(b) Solve by using the Simplex algorithm.

(c) Explain the solution—including why 20 acres are not being used and why only one product is grown—in terms the cooperative's members could understand.

First define the decision variables to be x_1 = number of acres planted with tomatoes, x_2 = number of acres planted with lettuce and x_3 = number of acres planted with squash. Then the objective function could be written

$$\text{Maximize P} = \frac{\$0.19}{\text{lb.}}\left(\frac{2000 \text{ lbs.}}{\text{acre}}\right) \cdot x_1 + \frac{\$0.10}{\text{lb.}}\left(\frac{3000 \text{ lbs.}}{\text{acre}}\right) \cdot x_2$$

$$+ \frac{\$0.25}{\text{lb.}}\left(\frac{1000 \text{ lbs.}}{\text{acre}}\right) \cdot x_3 - \frac{\$0.10}{\text{lb.}}\left(\frac{100 \text{ lbs.}}{\text{acre}}\right) \cdot x_1 - \frac{\$0.10}{\text{lb.}}\left(\frac{100 \text{ lbs.}}{\text{acre}}\right) \cdot x_2$$

$$- \frac{\$0.10}{\text{lb.}}\left(\frac{50 \text{ lbs.}}{\text{acre}}\right) \cdot x_3 - \frac{5 \text{ man-days}}{\text{acre}}\left(\frac{\$20}{\text{man-day}}\right) \cdot x_1$$

$$- \frac{6 \text{ man-days}}{\text{acre}}\left(\frac{\$20}{\text{man-day}}\right) \cdot x_2 - \frac{5 \text{ man-days}}{\text{acre}}\left(\frac{\$20}{\text{man-day}}\right) \cdot x_3$$

Subject to $5x_1 + 6x_2 + 5x_3 \leqslant 400$ (Manpower constraint)

$x_1 + x_2 + x_3 \leqslant 100$ (Land constraint)

$x_j \geqslant 0$ (Non-negative variables)

The objective function can be simplified to $P = 270x_1 + 170x_2 + 145x_3$. After adding slack variables to the constraints, our first tableau is

First Tableau:

c_i	BASIS	V_1	V_2	V_3	V_4	V_5	b_i	b_i/a_{ij}
0	x_4	⑤	6	5	1	0	400	80 →
0	x_5	1	1	1	0	1	100	100
	c_j	270	170	145	0	0	—	
	ΔOF_j	270	170	145	0	0	0	

Example 1 **123**

The objective function can be improved most by bringing x_1 into the basis to replace x_4. To transform the pivotal column into a $[1 \ 0]^T$ vector, we use the Gauss-Jordan elimination procedure and we obtain the second tableau:

Second Tableau:

c_i	BASIS	V_1	V_2	V_3	V_4	V_5	b_i	b_i/a_{ij}
270	x_1	1	$\frac{6}{5}$	1	$\frac{1}{5}$	0	80	
0	x_5	0	$-\frac{1}{5}$	0	$-\frac{1}{5}$	1	20	
	c_j	270	170	145	0	0	—	
	ΔOF_j	0	-154	-125	-54	0	21,600	

None of the ΔOF_j is positive for a nonbasic variable. Therefore, the optimal solution is to plant 80 acres with tomatoes (x_1) and to let 20 acres of land remain unused (x_5). The profit from the 80 acres of tomatoes is \$21,600.

The presence of a slack variable in the optimum solution means that we could make a higher profit from leaving 20 acres unplanted rather than attempting to grow lettuce or squash on it. In other words if the extra 20 acres were farmed, available labor would be diverted from tomatoes to something else and as a result profit would be reduced. With 80 acres of tomatoes planted, all 400 man-hours of available labor would be utilized.

In recognition of the uncertain nature of future market conditions, suppose we want to know how high the price of squash would have to be before we would consider growing it. From the second tableau, the ΔOF_3 for x_3 (squash) is -125. To get this value, recall that we performed this calculation: $145 - 1(270) - 0(0) = -125$. You can see that x_3 would enter the basis if ΔOF_3 were positive. This would happen when c_3 is greater than $145 + 125 = 270$, say 271. Now we can calculate the required price per pound for each acre of squash to be:

$$271 = \frac{1000 \text{ lbs. squash}}{\text{acre}} (\frac{\text{price}}{\text{lb.}}) - \frac{\$0.10}{\text{lb.}} (\frac{50 \text{ lbs.}}{\text{acre}}) - \frac{5 \text{ man-days}}{\text{acre}} (\frac{\$20}{\text{man-day}})$$

or

$$271 = 1000 (\frac{\text{price}}{\text{lb.}}) - 105$$

and

$$\frac{\text{price}}{\text{lb.}} = \underline{\underline{\$0.376}}$$

Thus, the price would have to increase from $0.25/lb. to $0.376/lb. before we find it profitable to grow squash.

EXAMPLE 2

A company has two grades of inspectors, A and B, the classification depending on the speed and accuracy of the inspector. Grade A calls for an inspection speed of 40 pieces per hour, with an accuracy of 95 percent and carries a wage rate of $4.00 per hour. Grade B requires a speed of 30 pieces per hour, with an accuracy of 90 percent and pays $2.25 per hour. It is required that a minimum of 2600 pieces be inspected per eight-hour day. There are seven Grade A inspectors and ten Grade B inspectors available for this inspection job. Each inspection error is estimated to cost $1.50. Find the "best" utilization of inspectors.

Let x_1 = number of Grade A inspectors and x_2 = number of Grade B inspectors. Then the objective function would be to minimize

$$C = \frac{\$4.00}{\text{hr.}}(x_1) + \frac{\$2.25}{\text{hr.}}(x_2) + \frac{2 \text{ defects}}{\text{hr.}}(\frac{\$1.50}{\text{defect}})(x_1) + \frac{3 \text{ defects}}{\text{hr.}}(\frac{\$1.50}{\text{defect}})(x_2)$$

or

$$\text{Minimize } C = 7x_1 + 6.75x_2$$

Subject to these constraints:

$$40x_1 + 30x_2 \geqslant 325 \quad \text{(a minimum of 325 pieces inspected/hour)}$$

$$x_1 \leqslant 7$$

$$x_2 \leqslant 10$$

$$x_1, x_2 \geqslant 0$$

Example 2 **125**

After adding surplus variables and artificial variables, the first tableau would be:

c_i	BASIS	V_1	V_2	V_3	V_4	V_5	V_6	b_i	b_i/a_{ij}
M	x_4	40	30	-1	1	0	0	325	$6\,5/8$
0	x_5	(1)	0	0	0	1	0	7	7 →
0	x_6	0	1	0	0	0	1	10	∞
	c_j	7	6.75	0	M	0	0	–	
	$\Delta 0F_j$	7	6.75	M	0	0	0	325M	
		-40M	-30M						
		↑							

By choosing nonbasic variable x_1 to enter the basis (it has the most negative $\Delta 0F_j$), it is apparent that surplus variable x_5 would leave. We next transform the coefficient matrix such that the pivotal column becomes $[0\ 1\ 0]^T$:

c_i	BASIS	V_1	V_2	V_3	V_4	V_5	V_6	b_i	b_i/a_{ij}
M	x_4	0	(30)	-1	1	-40	0	45	$3/2$ →
7	x_1	1	0	0	0	1	0	7	∞
0	x_6	0	1	0	0	0	1	10	10
	c_j	7	6.75	0	M	0	0	–	
	$\Delta 0F_j$	0	6.75	M	0	-7	0	49	
			-30M			+40M		+45M	
			↑						

Now we bring x_2 into the basis and remove the artificial variable, x_4, to obtain the third tableau.

c_i	BASIS	V_1	V_2	V_3	V_4	V_5	V_6	b_i	b_i/a_{ij}
6.75	x_2	0	1	$-1/30$		$-4/3$	0	$3/2$	
7	x_1	1	0	0		1	0	7	
0	x_6	0	0	$1/30$		$4/3$	1	$1\,7/2$	
	c_j	7	6.75	0		0	0	—	
	ΔOF_j	0	0	$9/40$		2	0	$\boxed{59\,1/8}$	

All ΔOF_j are positive, so the optimal solution is $x_1 = 7$, $x_2 = 3/2$ and $x_6 = 17/2$. The minimum cost is \$59 1/8 per hour. In a real-world problem we would want integer valued x_1 and x_2. A method for determining an optimal solution in integers will be described in the following chapter.

THE DUAL OF A

LINEAR

PROGRAMMING

PROBLEM

To this point we have been dealing with solutions to "primal" linear programming problems, and an optimal solution to the primal problem can be determined by using the Simplex Method. Associated with each primal formulation is another linear programming problem known as the "dual." Solving dual problems is often easier than solving primal problems when (a) the dual has fewer linear constraints (time required to solve linear programming problems is directly affected by the number of constraints)* and (b) the dual involves maximization of a linear objective function (it may be possible to avoid artificial variables that otherwise would be used in a primal minimization problem). As implied by (b), we shall be maximizing the objective function of the dual problem if the primal problem calls for minimizing an objective function, and vice-versa. Hence, for large linear programming problems it is frequently advantageous to work with the dual problem.

The dual of a linear programming problem can be constructed by defining a new decision variable for each constraint in the primal problem and a new

*The number of iterations necessary to converge on an optimum solution in the Simplex Method usually ranges from 1.5 to 3 times the number of structural constraints in the problem.

linear constraint for each variable in the primal. This concept is best illustrated in terms of an actual problem, so let us consider the primal problem formulated earlier on page 107:

$$\text{Maximize P} = 4x_1 + 3x_2$$

$$\text{Subject to} \quad x_1 + 2x_2/3 \leqslant 6000$$

$$x_1 \quad \leqslant 4000$$

$$x_2 \leqslant 6000$$

$$x_1, x_2 \quad \geqslant 0$$

The dual linear programming problem is

$$\text{Minimize P}' = 6000\, y_1 + 4000\, y_2 + 6000\, y_3$$

$$\text{Subject to} \quad y_1 + y_2 \quad \geqslant 4$$

$$2y_1/3 \quad + y_3 \quad \geqslant 3$$

$$y_1, y_2, y_3 \quad \geqslant 0$$

Here you can see that new dual variables (y_1, y_2 and y_3) have been defined, one for each constraint in the primal. Because there are two decision variables in the primal problem, we will have two constraints in the dual. Also, the coefficient of the jth variable in the dual's objective function is the ith component of the primal requirements vector. Furthermore, the dual's "requirements" vector consists of coefficients of decision variables in the primal objective function.

Coefficients of each constraint in the dual (i.e. row vectors) are the column vectors associated with each decision variable in the coefficient matrix of the primal problem. In other words, the coefficient matrix of the dual is the transpose of the primal's coefficient matrix. Finally, notice that maximizing the primal problem is equivalent to minimizing the dual. In this regard it can be shown that the optimum minimum solution to the dual equals the optimum maximum solution to the primal problem and vice-versa.

When a primal constraint is a "less than or equal" ($\leqslant$) inequality, the corresponding variable in the dual is non-negative. Furthermore, an equality in the primal problem means that the corresponding dual variable is unrestricted in sign. In summary form, the following relationships exist between constraints and variables in the primal and the dual.

Primal (maximization)   *Dual (minimization)*

ith inequality ($\leqslant$) $\longrightarrow$ $y_i \geqslant 0$

ith equality (=) $\longrightarrow$ y_i are unrestricted in sign

$x_j \geqslant 0$ $\longrightarrow$ jth inequality ($\geqslant$)

x_j are unrestricted in sign $\longrightarrow$ jth equality (=)

From these relationships you can see why we have "$\leqslant$" constraints in the primal and "$\geqslant$" constraints in the dual. Before solving the dual problem that is discussed above, it should be observed that the dual provides information regarding the marginal value ("shadow prices") of resources that are inputs to the system being optimized. A solution to the dual presents some interesting economic interpretations of linear programming problems in view of these transformations from the primal to the dual (compare formulations on page 127):

	Primal	Dual
Objective Function	Maximize P	Minimize P'
Number of Decision Variables	2	3
Number of Constraints	3	2
Constraint Constants (b_i)	Resources Available	Profit Per Unit of Output
Constraint Coefficients (a_{ij})	Inputs Required Per Unit of Output	Implicit Value of Resources

The optimum value of a dual decision variable indicates the rate of change in the objective function's value if resources associated with the corresponding *primal* constraint could be increased. For example, if $y_1 = 4$ in the optimal dual solution it would be possible to increase the primal objective function by four units for each *unit* increase in available resources in the first primal constraint. For this reason, the optimum value of the ith variable in the dual reflects the marginal value of the ith resource in the primal. When x_1, the first primal decision variable, does not occur in the optimal primal solution, the first dual slack variable *will* occur in the optimal dual solution.

To solve the dual problem by the Simplex Method, we would add slack (or surplus) variables where needed and possibly artificial variables to obtain an initial basic feasible solution. In this problem we already have a 2 × 2 identity matrix to use as a starting point, but surplus variables are still necessary to convert the inequalities to equalities as follows:

$$\text{Minimize } P' = 6000 y_1 + 4000 y_2 + 6000 y_3 + 0 y_4 + 0 y_5$$

$$\text{Subject to} \quad y_1 + y_2 - y_4 = 4$$

$$2y_1/3 + y_3 - y_5 = 3$$

$$y_i \geqslant 0$$

The first tableau is shown below.

First Tableau:

c_i	BASIS	V_1	V_2	V_3	V_4	V_5	b_i	b_i/a_{ij}
4000	y_2	①	1	0	-1	0	4	$4 \longrightarrow$
6000	y_3	$\frac{2}{3}$	0	1	0	-1	3	$\frac{9}{2}$
	c_j	6000	4000	6000	0	0	—	
	ΔOF_j	-2000	0	0	4000	6000	34,000	

$\uparrow$

Because we are minimizing the objective function, an optimal solution will be found when all ΔOF_j are positive for nonbasic variables. The y_j with the most negative ΔOF_j is brought into the solution, so here we introduce y_1 into the basis. Notice that the objective function assumes lower values from one iteration to another in minimization problems. The second tableau is:

Second Tableau:

c_i	BASIS	V_1	V_2	V_3	V_4	V_5	b_i	b_i/a_{ij}
6000	y_1	1	1	0	-1	0	4	
6000	y_3	0	$-\frac{2}{3}$	1	$\frac{2}{3}$	-1	$\frac{1}{3}$	
	c_j	6000	4000	6000	0	0	—	
	ΔOF_j	0	2000	0	2000	6000	26,000	

From the second tableau the optimal solution is seen to be $y_1 = 4$, $y_3 = 1/3$ and $P' = 26{,}000$. In comparing this solution with the primal solution discussed on page 115, it is apparent that $MAX. P = MIN. P'$. But what meaning can be attributed to the slack variables and decision variables in the optimal dual solution?

The dual slack variables have ΔOF_i values that correspond to b_i values of primal decision variables. The ΔOF_j of the first *slack* variable in the dual equals b_i of the first *decision* variable in the primal (i.e. $\Delta OF_4 = 2000$). Similarly, $\Delta OF_5 = 6000$ for the second slack variable in the dual, which equals b_2 for the second decision variable in the primal. Also observe that ΔOF_2 in the dual equals b_4 in the primal ($\Delta OF_2 = 2000$). The dual decision variables have ΔOF_j values that correspond to b_i values of primal slack variables. The b_i of the first decision variable in the dual, y_1, is identical to the $-\Delta OF_j$ of the first slack variable in the primal, x_3 and so forth.

Let us next compare the optimal primal tableau (page 115) with the optimal dual tableau (page 129). It can be seen that when a dual slack variable is in the optimum dual solution, the corresponding primal decision variable will not be in the optimal primal solution and vice-versa. Thus, by noting that y_4 and y_5 (first and second dual slack variables) are not in the optimal dual solution, we could infer that x_1 and x_2 (first and second primal decision variables) would be in the optimal primal solution. If there had been a *nonzero* dual slack variable in the optimum solution, the value of resources used in producing the associated primal decision variable would exceed the profit (or gain) derived from this variable.

In contrast, a zero-valued primal slack variable means that the corresponding dual decision variable is used to its maximum capacity. Thus, if the first slack variable in the optimal primal solution is zero, the first dual variable, y_1, is in the optimal dual solution since the first resource has a favorable marginal contribution to the objective function.

From the optimal dual solution, we see that the marginal value of resources in the first and third primal constraints is 4 and 1/3, respectively. The primal objective function could be increased by these amounts for each additional unit of resources made available in the first and third constraints (up to a limit, of course!). As you can now see, we do not have to solve the dual problem just to obtain the optimal values of the dual variables; they can be determined directly from the primal solution. The main reason for dealing with the dual is to take advantage of its computational simplicity.

As a final illustration of transforming the primal into the dual, we will set up this problem:

$$\text{Maximize } P = 3x_1 - 2x_2$$

$$\text{Subject to} \quad x_1 + x_2 \leqslant 9 \tag{1}$$

$$-2x_1 + x_2 \leqslant 4 \tag{2}$$

$$4x_1 + 3x_2 \geqslant 11 \tag{3}$$

$$x_1, \quad x_2 \geqslant 0$$

To avoid "mixed" constraints, we multiply the third structural constraint by -1 to obtain $-4x_1 - 3x_2 \leqslant -11$. Now that the maximization problem has all "$\leqslant$" constraints, the corresponding dual variables will be greater than or equal to zero and we can write the dual:

Minimize $P' = 9y_1 + 4y_2 - 11y_3$

Subject to $y_1 - 2y_2 - 4y_3 \geqslant 3$ (1')

$y_1 + y_2 - 3y_3 \geqslant -2$ (2')

$y_1, y_2, y_3 \geqslant 0$

After adding the appropriate surplus and artificial variables (i.e. y_5) and multiplying the second constraint by -1, the constraints would be

$$y_1 - 2y_2 - 4y_3 - y_4 + y_5 = 3 \tag{1'}$$

$$-y_1 - y_2 + 3y_3 + y_6 = 2 \tag{2'}$$

The Simplex Method would next be used to determine an optimal solution.

TWO COMPUTER
PROGRAMS FOR
SOLVING LINEAR
PROGRAMMING
PROBLEMS

As we have seen in Chapters 4 and 5, linear programming is a useful tool in determining the optimum solution to certain types of resource allocation problems. However, solving these problems by a graphical method (for problems dealing with only two variables) or manually performing the Simplex algorithm is both tedious and time consuming. For this reason, most computer centers have "canned" linear programming programs in their libraries.

a conversational
mode LP program

The FORTRAN IV program listed in Figure 5-1 solves linear programming problems by using the basic Simplex Method described in Chapter 5.* It can also be used to solve special types of linear programming problems to be discussed in Chapter 6 (all except for integer programming). This particular program can handle up to forty-nine (49) variables and forty-nine (49) constraints and is used at the University of Tennessee at Knoxville with its I.B.M. 360 (Model 65) Call O.S. system. The data is fed into the system by a remote terminal. The code name for the program is "L.P.".

Procedure

1. After loading program "L.P." into the computer, the operator types RUN.

2. The terminal will then "ask" if the initial basic feasible solution (B.F.S.) is available. If the initial basic feasible solution is known, the operator types "1." The computer then asks for this solution and its tableau. If the operator does not know the initial B.F.S., he types "2."

3. The computer next "asks" for the following data in order:

   (a) The number of variables and constraints

   (b) Whether the objective function is to be maximized or minimized

   (c) The cost vector (coefficients of objective function)

   (d) The coefficients of the constraint equations

   (e) The b_i vector (i.e., the requirements vector)

   (f) The equality or inequality conditions.

4. The program will add slack and artificial variables as needed and calculate its own basic feasible solution. The linear programming problem is then solved by the computer. The optimum value of the objective function and the values of the decision variables are printed by the computer, and if slack or surplus variables are in the solution, the corresponding constraint equations are printed.

*Reprinted by permission of D.H. Pike, Associate Professor of Industrial Engineering, University of Tennessee, Knoxville.

A few example problems and their computer printouts are now given to illustrate the use of the program. These problems are from the exercises at the end of Chapter 5—problems 7(a), 8(a), and 9(a), respectively.

Example 1 [Problem 7(a)] :

$$\text{Maximize } F(x) = 4x_1 - 2x_2 - 3x_3$$

$$\text{Subject to } \quad 2x_1 + x_2 \leqslant 2$$

$$-2x_1 + x_2 + x_3 \leqslant 1$$

$$4x_1 + 3x_2 \leqslant 5$$

$$\text{All } x_i \geqslant 0$$

*One must be sure to multiply any constraint equation with a negative b_i value by minus one (-1) before entering it in the program (see Example 3).

Referring to the printout shown in Figure 5-2, the operator types the appropriate data as shown. Notice that the first and third constraint equations do not have a variable x_3 and thus a zero is used as the coefficient in 3(d) above. We see that the solution contains two slack variables (x_5 and x_6) corresponding to constraint equations 2 and 3, respectively. Also notice that no values are given for x_2 and x_3. These variables are equal to zero in the optimal solution.

Example 2 [Problem 8(a)] :

$$\text{Minimize } P = -12x_1 - 5x_2$$

$$\text{Subject to } \quad 5x_1 + x_2 \leqslant 1500$$

$$2x_1 + 3x_2 \leqslant 1000$$

$$3x_1 + 2x_2 \leqslant 800$$

$$x_1 \geqslant 40$$

$$x_2 \geqslant 30$$

This problem is similar to the previous example except that it is a minimization problem. Also, it involves mixed inequalities in the constraints. The printout for this problem is shown in Figure 5-3.

Example 3 [Problem 9(a)] :

$$\text{Maximize } Z = 3x_2$$

$$\text{Subject to } 3x_1 + 2x_2 \leqslant 7$$

$$x_1 - x_2 \geqslant -2$$

$$x_1, x_2 \geqslant 0$$

Example 3 has one important difference from the two previous examples that the analyst must be alert to catch. Notice that the b_i vector has a negative element, and thus the second constraint equation must be multiplied by negative one (-1) before solving. All resources were used in this problem, and hence there are no slack variables in the final tableau. A computer printout of the optimal solution is shown in Figure 5-4.

FIGURE 5-1

```
5   DIMENSIØN INEQ(50),LART(50),TCØST(50),BCØST(50)
7   DIMENSIØN Y(50,50),K(50),YNEW(50,50),CØST(50),ISLAK(50)
20  WRITE(6,500)
30  500 FØRMAT(' IF INITIAL B.F.S. IS AVAILABLE TYPE 1; ØTHERWISE TYPE 2')
40  READ (5,*) IPRØG
50  IF(IPRØG.EQ.2)GØ TØ 11
60  WRITE(6,1000)
70  1000 FØRMAT(' ENTER NUMBER ØF VARIABLES AND CØNSTRAINTS')
80  READ (5,*) N,M
90  NP1=N+1
100 MP1=M+1
110 WRITE (6,2000)
120 2000 FØRMAT(' ENTER THE SUBSCRIPT ØF THE BASIC VARIABLES'/,%
121 ' IN PRØPER ØRDER FØLLØWING THE ?')
130 READ (5,*) (K(I),I=1,M)
140 WRITE (6,3000)
150 3000 FØRMAT(' ENTER TABLEAU RØWS WITH VALUES IN THE'/,%
151 ' FØLLØWING ØRDER ( B(I),X(1),X(2),X(3),ETC.)')
160 READ (5,*) ((Y(I,J),J=1,NP1),I=1,M)
170 WRITE(6,3100)
180 3100 FØRMAT(' TYPE 1 FØR MINIMIZE; TYPE 2 FØR MAXIMIZE')
190 READ (5,*) ITYPE
200 WRITE(6,1300)
```

(continued)

```
220  READ (5,*) (CØST(J),J=2,NP1)
230  CØST(1) = 0.0
240  IF(ITYPE.EQ.1)GØ TØ 3
250  DØ 17 J=1,NP1
260  17 CØST(J)=-CØST(J)
270  3 DØ 16 J=1,NP1
280  W=0.0
290  DØ 15 I=1,M
300  Z=CØST(K(I)+1)*Y(I,J)
310  15 W=Z+W
320  16 Y(MP1,J)=W-CØST(J)
325  PHASE = 2.0
330  CALL SØLV(Y,K,CØST,N,M,KUNB,PHASE)
335  IF(ITYPE.EQ.1)GØ TØ 19
336  DØ 18 J=1,NP1
337  18 Y(MP1,J)=-Y(MP1,J)
340  19 IF(KUNB.EQ.1)GØ TØ 10
345  WRITE(6,4101)
350  WRITE(6,4000) ((K(I),Y(I,1)),I=1,M)
360  4000 FØRMAT(1HO, ' X(',I2,') = ',F9.3)
370  WRITE(6,4500) Y(MP1,1)
380  4500 FØRMAT(1HO, ' Z=',F9.3)
385  GØ TØ 340
390  10 WRITE(6,5000)
400  5000 FØRMAT(' UNBØUNDED SØLUTIØN')
410  GØ TØ 340
1000  11 WRITE(6,1000)
1020  READ (5,*) N,M
1030  NP1=N+1
1035  N1=N+1
1036  M1=M
1040  MP1=M+1
1045  LD=0
1050  WRITE(6,3100)
1070  READ (5,*) ITYPE
1080  WRITE(6,1300)
1090  1300 FØRMAT(' ENTER THE CØST VECTØR FØLLØWING THE ?')
1100  READ (5,*) (BCØST(J),J=2,NP1)
1110  BCØST(1)=0.0
1120  WRITE(6,1400)
1130  1400 FØRMAT(' ENTER CØEFFICIENTS ØF CØNSTRAINT EQUATIØNS'/,%
1131  ' BY RØWS FØLLØWING THE ?')
1140  READ (5,*)((Y(I,J),J=2,NP1),I=1,M)
```
(continued)

```
1150  WRITE (6,1500)
1160  1500 FØRMAT(' ENTER B VECTØR FÓLLØWING THE ?')
1170  READ (5,*)(Y(I,1),I=1,M)
1180  WRITE (6,1600)
1190  1600 FØRMAT(' ENTER EQUALITY ØR INEQUALITY ',%
1191  ' CØNDITIØNS FØLLØWING THE ?',/' (E.G., .LT.=1, .EQ.=2',%
1192  ' .GT.=3)')
1200  READ (5,*) (INEQ(I),I=1,M)
1210  DØ 12 J=1,NP1
1220  12 TCØST(J)=0.0
1240  DØ 13 JJ=1,50
1250  LART(JJ)=0
1255  13 ISLAK(JJ)=0
1260  JJ=NP1+1
1270  DØ 105 I=1,M
1280  IF(INEQ(I).EQ.1)GØ TØ 81
1290  IF(INEQ(I).EQ.2)GØ TØ 105
1300  DØ 61 II=1,M
1310  61 Y(II,JJ)=0.0
1320  Y(I,JJ)=-1.0
1330  TCØST(JJ)=0.0
1340  ISLAK(JJ-1)=I
1410  JJ=JJ+1
1420  GØ TØ 105
1430  81 DØ 86 II=1,M
1440  86 Y(II,JJ)=0.0
1450  Y(I,JJ)=1.0
1460  TCØST(JJ)=0.0
1480  K(I)=JJ-1
1485  LC=LD
1486  LD=LC+1
1487  ISLAK(JJ-1)=I
1490  JJ=JJ+1
1569  105 CØNTINUE
1570  JJJ=N1
1571  DØ 109 I=1,M
1572  IF(INEQ(I).EQ.1)GØ TØ 109
1573  DØ 106 II=1,M
1574  106 Y(II,JJ)=0.0
1575  Y(I,JJ)=1.0
1576  TCØST(JJ)=1.0
1578  K(I)=JJ-1
1579  LART(JJ-1)=1
```

(continued)

```
1580  JJ=JJ+1
1581  109 CØNTINUE
1590  N=JJ-2
1600  NP1=N+1
1610  DØ 110 J=1,NP1
1620  110 CØST(J)=TCØST(J)
1630  DØ 130 J=1,NP1
1640  W=0.0
1650  DØ 120 I=1,M
1660  Z=CØST(K(I)+1)*Y(I,J)
1670  120 W=Z+W
1680  130 Y(MP1,J)=W-CØST(J)
1685  PHASE = 1.0
1690  CALL SØLV(Y,K,CØST,N,M,KUNB,PHASE)
1700  IF(Y(MP1,1).LE.0.01)GØ TØ 160
1710  WRITE(6,1700)
1720  1700 FØRMAT(' UNFEASIBLE SØLUTIØN')
1725  GØTØ340
1730  160 DØ 170 J=1,N1
1740  170 CØST(J)=BCØST(J)
1741  IF(ITYPE.EQ.1)GØ TØ 175
1742  DØ 200 J=1,NP1
1743  200 CØST(J)=CØST(J)
1750  175 DØ 190 J=1,NP1
1760  W=0.0
1770  DØ 180 I=1,M
1780  Z=CØST(K(I)+1)*Y(I,J)
1790  180 W=Z+W
1800  190 Y(MP1,J)=W-CØST(J)
1840  210 JJJ=N-M1+LD
1850  DØ 300 L=1,M
1860  IF(LART(K(L)).EQ.0)GØ TØ 300
1870  IRØWR=L
1880  213 KCØL=JJJ
1881  JJJJ=JJJ-1
1882  DØ 215 I=1,M
1883  IF(K(I).EQ.JJJJ)GØ TØ 217
1884  IF(Y(IRØWR,KCØL).EQ.0.0)GØ TØ 217
1885  215 CØNTINUE
1886  GØ TØ 214
1887  217 JJJ=JJJ-1
1888  IF(KCØL.GT.1)GØ TØ 213
1889  DØ 667 I=IRØWR,M
```

(continued)

```
1890 DØ 667 J=1,NP1
1891 K(I)=K(I+1)
1892 667 Y(I,J)=Y(I+1,J)
1893 M=M-1
1894 MP1=MP1-1
1895 GØ TØ 210
1896 214 JJJ=JJJ-1
1900 DØ 220 I=1,MP1
1910 DØ 220 J=1,NP1
1920 220 YNEW(I,J)=Y(I,J)-Y(I,KCØL)*Y(IRØWR,J)/Y(IRØWR,KCØL)
1925 DØ 230 J=1,NP1
1926 230 YNEW(IRØWR,J)=Y(IRØWR,J)/Y(IRØWR,KCØL)
1930 K(IRØWR)=KCØL-1
1940 DØ 240 I=1,MP1
1950 DØ 240 J=1,NP1
1955 240 Y(I,J)=YNEW(I,J)
1960 300 CØNTINUE
1961 PHASE=2.0
1962 N=N-M1+LD
1970 CALL SØLV(Y,K,CØST,N,M,KUNB,PHASE)
1971 IF(ITYPE.EQ.1)GØ TØ 325
1972 DØ 320 J=1,NP1
1973 320 Y(MP1,J)=-Y(MP1,J)
1980 325 IF(KUNB.EQ.1)GØ TØ 330
1985 WRITE(6,4101)
1986 4101 FØRMAT(' SØLUTIØN***********************',///)
1990 DØ 239 I=1,M
1991 IF(ISLAK(K(I)).EQ.0)GØ TØ 328
1992 WRITE(6,4200) K(I),Y(I,1),ISLAK(K(I))
1993 4200 FØRMAT(1H0,' X(',I2,') = ',F9.3,5X,' VARIABLE',%
1994 ' FØR CØNSTRAINT NUMBER ',I2)
1995 GØ TØ 329
1996 328 WRITE(6,4100) K(I),Y(I,1)
1997 4100 FØRMAT(1H0,' X(',I2,') = ',F9.3)
2000 329 CØNTINUE
2010 WRITE(6,4600) Y(MP1,1)
2020 4600 FØRMAT(1H0,' Z=',F9.3)
2030 GØ TØ 240
2040 330 WRITE(6,5100)
2050 5100 FØRMAT(' UNBØUNDED SØLUTIØN')
2070 340 CALL EXIT
2080 END
5000 SUBRØUTINE SØLV(Y,K,CØST,N,M,KUNB,PHASE)
```

(continued)

```
5010  DIMENSIØN Y(50,50),K(50),YNEW(50,50),CØST(50)
5020  KUNB=0
5030  NP1=N+1
5040  MP1=M+1
5050  5 B=-1.0
5051  IF(PHASE.EQ.2.0)GØ TØ 6
5052  IF(Y(MP1,1).LE.0.0)GØ TØ 90
5060  6 DØ 30 J=2,NP1
5070  IF(Y(MP1,J).LE.0.0)GØ TØ 30
5080  A=Y(MP1,J)
5090  IF(A.LE.B)GØ TØ 30
6000  B=A
6010  KCØL=J
6020  30 CØNTINUE
6030  IF(B.EQ.-1.0)GØ TØ 90
6040  50 J=KCØL
6050  U=-1.0
6060  Q=0.0
6070  JU=2
6080  DØ 60 I=1,M
6090  IF(Y(I,J).LE.0.0)GØ TØ 60
6100  RØWCH=Y(I,1)/Y(I,J)
6110  KU=2*JU
6120  JU=4
6130  U=RØWCH
6140  IF(KU.EQ.4)GØ TØ 55
6150  IF(U.GE.Q)GØ TØ 60
6160  55 Q=U
6170  IRØWR=I
6180  60 CØNTINUE
6190  IF(U.EQ.-1.0)GØ TØ 70
6200  GØ TØ 80
6210  70 KUNB=1
6220  GØ TØ 90
6230  80 DØ 100 I=1,MP1
6240  DØ 100 J=1,NP1
6250  100 YNEW(I,J)=Y(I,J)-Y(I,KCØL)*Y(IRØWR,J)/Y(IRØWR,KCØL)
6260  DØ 84 J=1,NP1
6270  84 YNEW(IRØWR,J)=Y(IRØWR,J)/Y(IRØWR,KCØL)
6280  K(IRØWR)=KCØL-1
6290  DØ 85 I=1,MP1
6300  DØ 85 J=1,NP1
6310  85 Y(I,J)=YNEW(I,J)
```

(continued)

```
6320 GØ TØ 5
6340 90 CØNTINUE
6350 RETURN
6360 END
```

FIGURE 5-2

```
IF INITIAL B.F.S. IS AVAILABLE TYPE 1;ØTHERWISE TYPE 2
?2

ENTER NUMBER ØF VARIABLES AND CØNSTRAINTS
?3,3

TYPE 1 FØR MINIMIZE; TYPE 2 FØR MAXIMIZE
?2

ENTER THE CØST VECTØR FØLLØWING THE ?
?4,-2,-3

ENTER CØEFFICIENTS ØF CØNSTRAINT EQUATIØNS
BY RØWS FØLLØWING THE ?
?2,1,0

?-2,1,1

?4,3,0

ENTER B VECTØR FØLLØWING THE ?
?2,1,5

ENTER EQUALITY ØR INEQUALITY CØNDITIØNS FØLLØWING THE ?
(E.G., .LT.=1, .EQ.=2, .GT.=3)
?1,1,1
SØLUTIØN***********************

X( 1) =    1.000
X( 5) =    3.000    VARIABLE FØR CØNSTRAINT NUMBER  2
X( 6) =    1.000    VARIABLE FØR CØNSTRAINT NUMBER  3
Z=  4.000
TIME 7 SECS.
```

FIGURE 5-3

```
IF INITIAL B.F.S. IS AVAILABLE TYPE 1;ØTHERWISE TYPE 2
?2
```

(continued)

ENTER NUMBER ØF VARIABLES AND CØNSTRAINTS
?2,5

TYPE 1 FØR MINIMIZE; TYPE 2 FØR MAXIMIZE
?1

ENTER THE CØST VECTØR FØLLØWING THE ?
?-12,-5

ENTER CØEFFICIENTS ØF CØNSTRAINT EQUATIØNS
BY RØWS FØLLØWING THE ?
?5,1

?2,3

?3,2

?1,0

?0,1

ENTER B VECTØR FØLLØWING THE ?
?1500,1000,800,40,30

ENTER EQUALITY ØR INEQUALITY CØNDITIØNS FØLLØWING THE ?
(E.G., .LT.=1, .EQ.=2, .GT.=3)
?1,1,1,3,3

SØLUTIØN************************

X(3) = 236.667   VARIABLE FØR CØNSTRAINT NUMBER 1

X(4) = 416.667   VARIABLE FØR CØNSTRAINT NUMBER 2

X(6) = 206.667   VARIABLE FØR CØNSTRAINT NUMBER 4

X(1) = 246.667

X(2) =   30.000

Z=-3110.000
TIME 7 SECS.

FIGURE 5-4

IF INITIAL B.F.S. IS AVAILABLE TYPE 1;ØTHERWISE TYPE 2
?2

ENTER NUMBER ØF VARIABLES AND CØNSTRAINTS
?2,2

TYPE 1 FØR MINIMIZE; TYPE 2 FØR MAXIMIZE
?2

ENTER THE CØST VECTØR FØLLØWING THE ?
(continued)

?0,3

ENTER CØEFFICIENTS ØF CØNSTRAINT EQUATIØNS
BY RØWS FØLLØWING THE ?
?3,2

?–1,1

ENTER B VECTØR FØLLØWING THE ?
?7,2

ENTER EQUALITY ØR INEQUALITY CØNDITIØNS FØLLØWING THE ?
(E.G., .LT.=1, .EQ.=2, .GT.=3)
?1,1

SØLUTIØN***********************

X(1) = 0.600

X(2) = 2.600

Z= 7.800
TIME 8 SECS.

a card input LP
program

Attributes of the Program.   The FORTRAN program listed in Figure 5-5 solves linear programming problems via the Simplex algorithm.* It can solve problems of any dimension that will fit into core memory. It does not advise the user if his problem has no unique or feasible solution, nor does it contain a routine for dealing with degeneracy. Its printout shows the initial and successive tableaux (if the user asks for this added information) as well as the optimal solution. The user inputs his problem as a data deck following the program in Figure 5-5. Essentially, the data deck portrays the information in the tableau of the initial basic feasible solution.

The Data Deck.   To use the program, a data deck should be prepared as follows:

First Card.   The first card contains three data. These may be punched in whatever fashion the user desires, but Format Statement 104 must be changed

---

*Reprinted by permission of John Wiley & Sons, Inc. From McMillan, Claude, Jr. *Mathematical Programming: An Introduction to the Design and Application of Optimal Decision Machines.* New York: John Wiley & Sons, Inc., 1970. Appendix A.

accordingly. The variable names into which these three data are read and their purposes are as follows:

IW—The number of rows of constraint equations.

IZ—The number of columns in the initial matrix, including the column of constants in the requirements vector.

IY—The number of real variables + 1 (a real variable meaning variables in the initial tableau other than artificial and slack variables).

Second Card.   The second card contains only one datum: a "1" or a "0" in the first column. If the user wants the successive tableaux printed out as the iterative process progresses, he punches a "1"; otherwise he punches a "0" in column 1.

Third Card.   Onto the third card the user punches the coefficients of the objective function row and, if necessary, changes Format Statement 102 for reading this array into the subscripted variable P(N).

Fourth and Subsequent Cards.   Onto the next set of cards the user punches the coefficients in the initial tableau including values in the requirements (b_i) vector and, if necessary, changes Format Statement 103 in such a fashion that these data will be read into the subscripted variable D(M,N) as follows:

D(1,N)—Contains the coefficients (elements) in the first row of the tableau (thus the first constraint equation).

D(2,N)—Contains the coefficients (elements) in the second row of the tableau (thus the second constraint equation).

.

.

.

D(M,N)—Contains the coefficients (elements) in the Mth and final row of the tableau (thus the final constraint equation).

Furthermore:

D(M,1)—Should contain the first coefficient (element) in the Mth row.

D(M,2)–Should contain the second coefficient (element) in the Mth row.

.

.

.

D(M,N)–Should contain the Nth element in the Mth row, this being the value of b_i for the Mth constraint in the initial tableau.

Clearly, the dimensions of D(M,N) will be N = 1,IZ, M = 1,IW.

Finally, Format Statement 113 should be suitably composed for printing out the initial tableau, stored in D(M,N),N=1,IZ. In addition, the dimension statement should be adjusted (equipped with suitable arguments) for reserving space during execution as follows:

D(M,N)	where	$M \geqslant IW; N \geqslant IZ$
P(N)	where	$N \geqslant IZ - 1$
IBV(L)	where	$L \geqslant IW$
SC(J)	where	$J \geqslant IZ - 1$

FIGURE 5-5 The General Program Listing

```
 1        DIMENSION D(33,33),P(33),IBV(33),SC(33)
 2    101 FORMAT(I1)
 3    102 FORMAT((8F10.0)/(8F10.0)/(8F10.0)  )
 4    103 FORMAT((8F10.0)/(8F10.0)/(8F10.0)  )
 5    104 FORMAT(516)
 6    106 FORMAT(1H0,11HTABLEAU NO.,I6)
 7    108 FORMAT(1H1,8HSOLUTION)
 8    109 FORMAT(1H0,8HVARIABLE,4X,5HVALUE)
 9    110 FORMAT (1H ,2HX(,I3,4H) = ,F12.2)
10    111 FORMAT(1H0,27HALL OTHER VARIABLES = ZERO.)
11    112 FORMAT(1H1,20HTHE INITIAL TABLEAU.)
12    113 FORMAT (1H , 7X, 3X, 12F10.4/ (11X, 12F10.4))
13    300 FORMAT (8X,12I10)
14    301 FORMAT (1H0, 2X, 2HX(, I2, 1H), 3X, 12F10.3/(11X, 12F10.3))
15    302 FORMAT (11H0SIMPLEX CR, 12F10.3/ (11X, 12F10.3))
16    305 FORMAT (1H , 9H0BJ FNCTN, 1X,12F10.3/ (11X, 12F10.3))
17    789 FORMAT(1H0,10X,30HOBJECTIVE FUNCTION VALUE IS   ,F15.5)
```

(continued)

```
18        READ(5,104) IW,IZ,IY
19        IX=IZ-1
20        READ(5,101) ITAB
21        READ(5,102) (P(M) ,M=1,IX)
22        DO 15M=1,IW
23     15 READ(5,103) (D(M,N),N=1,IZ )
24        WRITE(6,112)
25        WRITE(6,300) (N,N=1,IX)
26        WRITE(6,305)(P(M),M=1,IX)
27        DO 16 M = 1,IW
28     16 WRITE(6,113) (D(M,N),N=1,IZ)
29        DO 20 N=IY,IX
30        DO 30 L=1,IW
31        IF(D(L,N).EQ.1.) GO TO 40
32     30 CONTINUE
33        GO TO 20
34     40 IBV(L)=N
35     20 CONTINUE
36        Z=0.
37        DO 210 M = 1,IW
38        IBVM = IBV(M)
39    210 Z=Z +D(M,IZ)* P(IBVM)
40        NOPIVS = 0
41        IF(ITAB .NE. 1) GO TO 13
42     13 SCMAX = 0
43        DO 31 N=1,IX
44        DO 32 I=1, IW
45        IF(N.EQ. IBV(I)) GO TO 31
46     32 CONTINUE
47        SUM = 0.
48        DO 33 I=1,IW
49        J=IBV(I)
50     33 SUM = SUM + P(J) * D(I,N)
51        SC(N) = P(N)-SUM
52        IF(SC(N).LE.SCMAX)GO TO 31
53        SCMAX = SC(N)
54        IPIVCO=N
55     31 CONTINUE
56        DO 200 M=1, IW
57        IBVM = IBV(M)
58    200 SC(IBVM) = 0.
59        IF(SCMAX.LE.0) GO TO 14
60        NOPIVS = NOPIVS+1
```

(continued)

```
61      3 SMLVAL = 999999
62        DO 4 M=1,IW
63        IF( D(M,IPIVCO)) 4, 4, 5
64      5 QUONT = D(M,IZ) / D(M,IPIVCO)
65        IF(QUONT - SMLVAL) 6,4,4
66      6 IPIVRO = M
67        SMLVAL = QUONT
68      4 CONTINUE
69        IBV(IPIVRO) = IPIVCO
70        DIV= D(IPIVRO,IPIVCO)
71        DO 7 N=1,IZ
72        CRANK = D(IPIVRO, N)
73      7 D(IPIVRO,N) = CRANK/DIV
74        IF(ITAB.NE.1) GO TO 12
75        WRITE(6,302) (SC(J),J=1,IX)
76        N100 = NOPIVS + 1
77        WRITE(6,789) Z
78        WRITE (6,106) N100
79        WRITE(6,300) (N,N=1,IX)
80     12 DO 10 M=1,IW
81        IF(M-IPIVRO)9,8,9
82      9 CM = -D(M,IPIVCO)
83        DO 11 N=1,IZ
84        TM = D(IPIVRO,N)* CM
85        SINKS = D(M,N)
86     11 D(M,N) = SINKS + TM
87      8 IF(ITAB.NE.1) GO TO 10
88        WRITE(6,301) IBV(M), (D(M,N) , N=1,IZ)
89     10 CONTINUE
90        Z = Z +SMLVAL * SCMAX
91        GO TO 13
92     14 WRITE(6,108)
93        WRITE(6,109)
94        DO 21 M= 1,IW
95     21 WRITE(6,110) IBV(M), D(M,IZ)
96        WRITE(6,111)
97        WRITE (6,789) Z
98        STOP
99        END
```

Example Problem 1:

Assume we are confronted with the following linear programming problem:

$$\text{Maximize } f = 20x_1 + 10x_2 + 5x_3$$

Subject to   (1) $5x_1 + 3x_2 + x_3 \leqslant 1050$

(2) $4x_1 + 3x_2 + 2x_3 \leqslant 1000$

(3) $x_1 + 2x_2 + 2x_3 \leqslant 400$

After adding slack variables x_4, x_5 and x_6, we obtain the initial tableau in Table 5-2.

TABLE 5-2

Objection function $\rightarrow$ 20	10	5	0	0	0	
x_1	x_2	x_3	x_4	x_5	x_6	Index, b_i
5	3	1	1	0	0	1050
4	3	2	0	1	0	1000
1	2	2	0	0	1	400

A suitable data deck prepared in conformance with the preceding instructions appears as follows:

Column number→6	12	18				
Card 1 3	7	4				

Column number 1						
Card 2 1						

Column number→10	20	30	40	50	60	70
Card 3 20.	10.	5.	0.	0.	0.	0.
Card 4 5.	3.	1.	1.	0.	0.	1050.
Card 5 4.	3.	2.	0.	1.	0.	1000.
Card 6 1.	2.	2.	0.	0.	1.	400.

This data deck is placed after the END statement in the program of Figure 5-5. When the program has been run on a computer, the output listing shown as Figure 5-6 is obtained.

FIGURE 5-6   Solution to Example Problem 1

THE INITIAL TABLEAU.

	1	2	3	4	5	6	
OBJ FNCTN	20.000	10.000	5.000	0.000	0.000	0.000	
	5.0000	3.0000	1.0000	1.0000	0.0000	0.0000	1050.0000
	4.0000	3.0000	2.0000	0.0000	1.0000	0.0000	1000.0000
	1.0000	2.0000	2.0000	0.0000	0.0000	1.0000	400.0000

SIMPLEX CR	20.000	10.000	5.000	0.0000	0.000	0.000	

OBJECTIVE FUNCTION VALUE IS   0.00000

TABLEAU NO.   2

	1	2	3	4	5	6	
X(1)	1.000	0.600	0.200	0.200	0.000	0.000	210.000
X(5)	0.000	0.600	1.200	-0.800	1.000	0.000	160.000
X(6)	0.000	1.400	1.800	-0.200	0.000	1.000	190.000

SIMPLEX CR	0.000	-2.000	1.000	-4.000	0.000	0.000	

OBJECTIVE FUNCTION VALUE IS   4200.00000

TABLEAU NO.   3

	1	2	3	4	5	6	
X(1)	1.000	0.444	0.000	0.222	0.000	-0.111	188.889
X(5)	0.000	-0.333	0.000	-0.667	1.000	-0.667	33.333
X(3)	0.000	0.778	1.000	-0.111	0.000	0.556	105.556

SOLUTION

VARIABLE		VALUE
X(1)	=	188.89
X(5)	=	33.33
X(3)	=	105.56

ALL OTHER VARIABLES = ZERO.

OBJECTIVE FUNCTION VALUE IS   4305.55400

Example Problem 2:

Assume we are confronted with the following linear programming problem:

$$\text{Minimize } P = -12x_1 - 5x_2$$

$$\text{Subject to} \quad 5x_1 + x_2 \leqslant 1500$$

$$2x_1 + 3x_2 \leqslant 1000$$

$$3x_1 + 2x_2 \leqslant 800$$

$$x_1 \geqslant 40$$

$$x_2 \geqslant 30$$

Changing P to a maximization problem and making all constraints "less than or equal to" produces this formulation:

$$\text{Maximize } P = 12x_1 + 5x_2$$

$$\text{Subject to} \quad 5x_1 + x_2 \leqslant 1500$$

$$2x_1 + 3x_2 \leqslant 1000$$

$$3x_1 + 2x_2 \leqslant 800$$

$$-x_1 \leqslant -40$$

$$-x_2 \leqslant -30$$

Adding slack and artificial variables we obtain the initial tableau in Table 5-3.

TABLE 5-3

Objective function→12	5	0	0	0	0	0	-1000	-1000	
x_1	x_2	x_3	x_4	x_5	x_6	x_7	x_8	x_9	Index, b_i
5	1	0	0	1	0	0	0	0	1500
2	3	0	0	0	1	0	0	0	1000
3	2	0	0	0	0	1	0	0	800
1	0	-1	0	0	0	0	1	0	40
0	1	0	-1	0	0	0	0	1	30

A suitable data deck prepared in conformance with the preceding instructions appears as follows:

Column number→6	12	18
Card 1 5	8	3

Column number→1
Card 2   1

Column number→8	16	24	32	40	48	56	64	72	80
Card 3 12.	5.	0.	0.	0.	0.	0.	-1000.	-1000.	0.
Card 4 5.	1.	0.	0.	1.	0.	0.	0.	0.	1500.
Card 5 2.	3.	0.	0.	0.	1.	0.	0.	0.	1000.
Card 6 3.	2.	0.	0.	0.	0.	1.	0.	0.	800.
Card 7 1.	0.	-1.	0.	0.	0.	0.	1.	0.	40.
Card 8 0.	1.	0.	-1.	0.	0.	0.	0.	1.	30.

The above data deck is entered following the END statement in the program of Figure 5-7. Here the program of Figure 5-5 has been reformatted for Example Problem 2. Then the FORTRAN program generates the output shown in Figure 5-8.

FIGURE 5-7

```
1        DIMENSION D(33,33),P(33),IBV(33),SC(33)
2    101 FORMAT(I1)
3    102 FORMAT (10F8.0)
4    103 FORMAT (10F8.0)
5    104 FORMAT(5I6)
6    106 FORMAT(1H0,11HTABLEAU NO.,I6)
7    108 FORMAT(1H1,8HSOLUTION)
8    109 FORMAT(1H0,8HVARIABLE,4X,5HVALUE)
9    110 FORMAT (1H ,2HX(,I3,4H) = ,F12.2)
10   111 FORMAT(1H0,27HALL OTHER VARIABLES = ZERO.)
11   112 FORMAT(1H1,20HTHE INITIAL TABLEAU.)
12   113 FORMAT (1H , 7X, 3X, 12F10.4/ (11X, 12F10.4))
13   300 FORMAT (8X,12I8)
14   301 FORMAT (1H0, 2X, 2HX(, I2, 1H), 3X, 12F10.3/(11X, 12F10.3))
15   302 FORMAT (11H0SIMPLEX CR, 12F10.3/ (11X, 12F10.3))
16   305 FORMAT (1H , 9HOBJ FNCTN, 1X,12F10.3/ (11X, 12F10.3))
17   789 FORMAT(1H0,10X,30HOBJECTIVE FUNCTION VALUE IS F15.5)
             F15.5)
```

```
18          READ(5,104) IW,IZ,IY
19          IX=IZ-1
20          READ(5,101) ITAB
21          READ(5,102) (P(M),M=1,IX)
22          DO 15M=1,IW
23      15  READ(5,103) (D(M,N),N=1,IZ )
24          WRITE(6,112)
25          WRITE(6,300) (N,N=1,IX)
26          WRITE(6,305)(P(M),M=1,IX)
27          DO 16 M = 1,IW
28      16  WRITE(6,113) (D(M,N),N=1,IZ)
29          DO 20 N=IY,IX
30          DO 30 L=1,IW
31          IF(D(L,N).EQ.1.) GO TO 40
32      30  CONTINUE
33          GO TO 20
34      40  IBV(L)=N
35      20  CONTINUE
36          Z=0.
37          DO 210 M = 1,IW
38          IBVM = IBV(M)
39     210  Z=Z +D(M,IZ)* P(IBVM)
40          NOPIVS = 0
41          IF(ITAB .NE. 1)GO TO 13
42      13  SCMAX = 0
43          DO 31 N=1,IX
44          DO 32 I=1,IW
45          IF(N.EQ. IBV(I)) GO TO 31
46      32  CONTINUE
47          SUM = 0.
48          DO 33 I=1,IW
49          J=IBV(I)
50      33  SUM = SUM + P(J) * D(I,N)
51          SC(N) = P(N)-SUM
52          IF(SC(N).LE.SCMAX)GO TO 31
53          SCMAX = SC(N)
54          IPIVCO=N
55      31  CONTINUE
56          DO 200 M=1, IW
57          IBVM = IBV(M)
58     200  SC(IBVM) = 0.
59          IF(SCMAX.LE.0) GO TO 14
60          NOPIVS = NOPIVS+1
```

```
61      3  SMLVAL = 999999
62         DO 4 M=1,IW
63         IF( D(M,IPIVCO)) 4, 4, 5
64      5  QUONT = D(M,IZ) / D(M,IPIVCO)
65         IF(QUONT - SMLVAL) 6,4,4
66      6  IPIVRO = M
67         SMLVAL = QUONT
68      4  CONTINUE
69         IBV(IPIVRO) = IPIVCO
70         DIV= D(IPIVRO,IPIVCO)
71         DO 7 N=1,IZ
72         CRANK = D(IPIVRO, N)
73      7  D(IPIVRO,N) = CRANK/DIV
74         IF(ITAB.NE.1) GO TO 12
75         WRITE (6,302) (SC(J),J=1,IX)
76         N100 = NOPIVS + 1
77         WRITE(6,789) Z
78         WRITE (6,106) N100
79         WRITE(6,300) (N,N=1,IX)
80     12  DO 10 M=1,IW
81         IF(M-IPIVRO)9,8,9
82      9  CM = -D(M,IPIVCO)
83         DO 11 N=1,IZ
84         TM = D(IPIVRO,N)* CM
85         SINKS = D(M,N)
86     11  D(M,N) = SINKS + TM
87      8  IF(ITAB.NE.1) GO TO 10
88         WRITE(6,301) IBV(M), (D(M,N) , N=1,IZ)
89     10  CONTINUE
90         Z = Z +SMLVAL * SCMAX
91         GO TO 13
92     14  WRITE(6,108)
93         WRITE(6,109)
94         DO 21 M= 1,IW
95     21  WRITE(6,110) IBV(M), D(M,IZ)
96         WRITE (6,111)
97         WRITE (6,789) Z
98         STOP
99         END
```

FIGURE 5-8

THE INITIAL TABLEAU.

	1	2	3	4	5	6	7	8	9	
OBJ FNCTN	12.000	5.000	0.000	0.000	0.000	0.000	0.000	-1000.000	-1000.000	
	5.0000	1.0000	0.0000	0.0000	1.0000	0.0000	0.0000	0.0000	0.0000	1500.0000
	2.0000	3.0000	0.0000	0.0000	0.0000	1.0000	0.0000	0.0000	0.0000	1000.0000
	3.0000	2.0000	0.0000	0.0000	0.0000	0.0000	1.0000	0.0000	0.0000	800.0000
	1.0000	0.0000	-1.0000	0.0000	0.0000	0.0000	0.0000	1.0000	0.0000	40.0000
	0.0000	1.0000	0.0000	-1.0000	0.0000	0.0000	0.0000	0.0000	1.0000	30.0000
SIMPLEX CR	1012.000	1005.000	-1000.000	-1000.000	0.0000	0.000	0.000	0.000	0.000	

OBJECTIVE FUNCTION VALUE IS -70000.00000

TABLEAU NO. 2

	1	2	3	4	5	6	7	8	9	
X(5)	0.000	1.000	5.000	0.000	1.000	0.000	0.000	-5.000	0.000	1300.000
X(6)	0.000	3.000	2.000	0.000	0.000	1.000	0.000	-2.000	0.000	920.000
X(7)	0.000	2.000	3.000	0.000	0.000	0.000	1.000	-3.000	0.000	680.000
X(1)	1.000	0.000	-1.000	0.000	0.000	0.000	0.000	1.000	0.000	40.000
X(9)	0.000	1.000	0.000	-1.000	0.000	0.000	0.000	0.000	1.000	30.000
SIMPLEX CR	0.000	1005.000	12.000	-1000.000	0.000	0.000	0.000	-1012.000	0.000	

OBJECTIVE FUNCTION VALUE IS -29520.00000

TABLEAU NO. 3

	1	2	3	4	5	6	7	8	9
X(5)	0.000	0.000	5.000	1.000	1.000	0.000	0.000	-5.000	1270.000
X(6)	0.000	0.000	2.000	3.000	0.000	1.000	0.000	-2.000	830.000
X(7)	0.000	0.000	3.000	2.000	0.000	0.000	1.000	-3.000	620.000
X(1)	1.000	0.000	-1.000	0.000	0.000	0.000	0.000	1.000	40.000
X(2)	0.000	1.000	0.000	-1.000	0.000	0.000	0.000	0.000	30.000
SIMPLEX CR	0.000	0.000	12.000	5.000	0.000	0.000	0.000	-1012.000	-1005.000

OBJECTIVE FUNCTION VALUE IS   630.00000

TABLEAU NO. 4

	1	2	3	4	5	6	7	8	9
X(5)	0.000	0.000	0.000	-2.333	1.000	0.000	-1.667	0.000	236.667
X(6)	0.000	0.000	0.000	1.667	0.000	1.000	-0.667	0.000	416.667
X(3)	0.000	0.000	1.000	0.667	0.000	0.000	0.333	-1.000	206.667
X(1)	1.000	0.000	0.000	0.667	0.000	0.000	0.333	0.000	246.667
X(2)	0.000	1.000	0.000	-1.000	0.000	0.000	0.000	0.000	30.000

FIGURE 5-8   (Continued)

SOLUTION

VARIABLE		VALUE
X(5)	=	236.67
X(6)	=	416.67
X(3)	=	206.67
X(1)	=	246.67
X(2)	=	30.00

ALL OTHER VARIABLES = ZERO.

OBJECTIVE FUNCTION VALUE IS   3109.99900

FIGURE 5-8   (Continued)

Articles included in the Appendix that illustrate the material of this chapter are:

"Linear Programming Guides Parts Buyers" by John L. Simpson and Dante J. Peller.

"Optimum Press Loadings in the Book Manufacturing Industry" by Charles H. Aikens III.

"Introduction to Linear Programming for Production" by J. Frank Sharp.

SUGGESTED ADDITIONAL READINGS

Smythe, W. R., and Johnson, L. A. *Introduction to Linear Programming with Applications.* Englewood Cliffs, N.J.: Prentice-Hall, Inc., 1966.

Strum, Jay E. *Introduction to Linear Programming.* San Francisco: Holden-Day, Inc., 1972.

Taha, H. A. *Operations Research, An Introduction.* New York: The Macmillan Company, 1971.

Dantzig, George B. *Linear Programming and Extensions.* Princeton, N.J.: Princeton University Press, 1963.

Cooper, L., and Steinberg, D. *Introduction to Methods of Optimization.* Philadelphia: W.B. Saunders Co., 1970.

Wagner, H. M. *Principles of Operations Research With Applications to Managerial Problems.* Englewood Cliffs, N.J.: Prentice-Hall, Inc., 1969.

EXERCISES

1. In your own words, define briefly these terms:

    (a) Basic solutions

(b)   Feasible region

(c)   Feasible solutions

(d)   Basic feasible solutions

(e)   Simplex criteria

(f)   Requirements vector

(g)   Convex region

(h)   Degeneracy

(i)   Artificial variable

(j)   Pivotal row

(k)   Pivotal element

(l)   Unbounded solution

(m)   Linear function

(n)   Decision variable

2.   Explain why the dual may be preferred to the primal when solving a linear programming problem.

3.   (a) In what situation would the Simplex Method not necessarily produce a unique solution to a linear programming problem? Draw a simple graph to illustrate your answer.

(b) When one or more artificial slack variables remain in the final Simplex tableau, the profit (or cost) coefficient in the objective function may not have been "undesirable" enough. What else might this situation imply?

(c) If all b_i/a_{ij} are zero or negative at a particular iteration, what can you say about the solution to a linear programming problem? If a b_i/a_{ij} ratio has a value of ∞, what does this mean?

4.   Give the *initial* tableau for a Simplex solution of the following problem.

$$\text{Maximize } P = 6x_1 + 3x_2 - 5x_3$$

$$\text{Subject to} \quad x_1 + x_2 + x_3 \leqslant 7$$

$$2x_1 \qquad + 7x_3 \geqslant 4$$

$$7x_1 \qquad + 6x_3 = 25$$

$$\text{All} \quad x_j \geqslant 0$$

5. When we have n decision variables in a linear programming problem and m inequality constraints, there are $\dfrac{(n+m)\,!}{n!\,m!}$ basic solutions (intersections) to evaluate. In the linear programming problem below, determine (a) all basic solutions, and (b) all basic feasible solutions.

$$\text{Maximize } P = 4w_1 + 3w_2$$

$$\begin{array}{lll} \text{Subject to} & 6w_1 + 4w_2 \leqslant 72 & w_1 \geqslant 0 \\ & 6w_1 + 7w_2 \leqslant 70 & w_2 \geqslant 0 \\ & 3w_1 + 10w_2 \leqslant 75 & \end{array}$$

6. The following Simplex tableau has been given to you by an associate who is confused about what to do next. He tells you that originally there were three slack variables, each with a coefficient of 0 in the objective function to be maximized. Show your associate how to complete the problem and interpret the results.

c_i	BASIS	V_1	V_2	V_3	V_4	V_5	b_i
3	x_2	0	1	2	0	-1	2
0	x_4	0	0	2	1	-1	1
1	x_1	1	0	-3	0	$3/2$	4

7. (a) Find the value of Z_1, Z_2 and Z_3 that solves this problem:

$$\text{Maximize } f = 4Z_1 - 2Z_2 - 3Z_3$$

$$\begin{array}{ll} \text{Subject to} & 2Z_1 + Z_2 \leqslant 2 \\ & -2Z_1 + Z_2 + Z_3 \leqslant 1 \\ & 4Z_1 + 3Z_2 \leqslant 5 \end{array}$$

$$\text{All } Z_i \geqslant 0$$

(b) Considering the above as the primal problem, state the form of the dual. Do not give a tableau, just state the dual in algebraic terms. From your final tableau of the primal problem above, what values of y_1, y_2 and y_3 will give the optimum solution to the dual?

8.   Consider the linear programming problem below:

$$\text{Minimize P}' = -12x_1 - 5x_2$$

$$\begin{aligned}
\text{Subject to} \quad & 5x_1 + x_2 \leqslant 1500 \\
& 2x_1 + 3x_2 \leqslant 1000 \\
& 3x_1 + 2x_2 \leqslant 800 \\
& x_1 \geqslant 40 \\
& x_2 \geqslant 30
\end{aligned}$$

(a)   Solve this problem by using the Simplex Method.

(b)   Formulate the dual problem and solve it.

(c)   Compare the number of basic solutions possible in part (a) and part (b).

9.   Find the optimal solution to these linear programming problems:

(a)

$$\text{Maximize Z} = 3x_2$$

$$\begin{aligned}
\text{Subject to} \quad & 3x_1 + 2x_2 \leqslant 7 \\
& x_1 - x_2 \geqslant -2 \\
& x_1, x_2 \geqslant 0
\end{aligned}$$

(b)

$$\text{Minimize C} = 3x_1 + 2x_2 + 4x_3 + 8x_4$$

$$\begin{aligned}
\text{Subject to} \quad & x_1 + 2x_2 + 5x_3 + 6x_4 \geqslant 8 \\
& -2x_1 + 5x_2 + 3x_3 - 5x_4 \leqslant 3 \\
& x_1, \quad x_2, \quad x_3, \quad x_4 \geqslant 0
\end{aligned}$$

(c)

$$\text{Maximize P} = 5x_1 + 3x_2$$

$$\text{Subject to} \quad 4x_1 + 5x_2 \leqslant 7$$

$$9x_1 + 2x_2 = 12$$

$$4x_1 + x_2 \leqslant 5$$

$$x_1, \quad x_2 \geqslant 0$$

10. (a) The following is the final tableau of a linear programming problem:

BASIS	V_1	V_2	V_3	V_4	V_5	V_6	b_i
x_5	2	4	1	0	1	0	30
x_4	1	-2	1	1	0	0	2
x_6	-2	3	-1	0	0	1	1

Maximize $P = \Sigma c_i b_i = 161$

The vectors V_1, V_2 and V_3 are slack variables which constituted the initial basis, and the objective function to be maximized was $P = 4x_4 + 5x_5 + 3x_6$. Reconstruct the initial tableau of this linear programming problem. *Hint*: In any tableau we can take the inverse of the vectors that were basic in the original coefficient matrix and multiply it by the other column vectors (including b_i) to obtain the original coefficients in these vectors.

(b) The optimal tableau for a linear programming problem is:

c_i BASIS	V_1	V_2	V_3	V_4	V_5	V_6	V_7	b_i
4 x_1	1	5/7	0	- 5/7	10/7	0	-1/7	50/7
0 x_6	0	-6/7	0	13/7	-61/7	1	4/7	325/7
9 x_3	0	2/7	1	12/7	- 3/7	0	1/7	55/7
ΔOF_j	0	-3/7	0	-11/7	-13/7	0	-5/7	695/7

Knowing only that V_5, V_6 and V_7 were in the original basis and that

$$[V_5 \ V_6 \ V_7]^{-1} = \begin{bmatrix} 1 & 0 & 1 \\ 7 & 1 & 3 \\ 3 & 0 & 10 \end{bmatrix}$$

verify that the original problem is:

$$\text{Maximize } P = 4x_1 + 5x_2 + 9x_3 + 11x_4$$

$$\text{Subject to} \quad x_1 + x_2 + x_3 + x_4 \leqslant 15$$

$$7x_1 + 5x_2 + 3x_3 + 2x_4 \leqslant 120$$

$$3x_1 + 5x_2 + 10x_3 + 15x_4 \leqslant 100$$

$$x_1, x_2, x_3, x_4 \geqslant 0$$

11. Shown below is a tableau of a linear programming maximization problem:

BASIS	V_1	V_2	V_3	V_4	V_5	b_i
x_3	1	0	1	0	0	4
x_2	0	1	0	1	0	3
x_5	1	0	0	-2	1	2
c_j	2	5	0	0	0	—

(a) Complete this tableau and use the Gauss-Jordan elimination procedure to determine the next tableau. If the optimum solution is obtained, state it along with the values of all decision variables and slack variables.

(b) Suppose in the original tableau that vectors V_2, V_3 and V_5 were the following:

$$V_2 = \begin{bmatrix} 0 \\ 1 \\ 2 \end{bmatrix} \qquad V_3 = \begin{bmatrix} 1 \\ 0 \\ 0 \end{bmatrix} \qquad V_5 = \begin{bmatrix} 0 \\ 0 \\ 1 \end{bmatrix}$$

Write the original constraints and the objective function.

12. The following linear programming problem is given:

$$\text{Minimize } Z = 2x_1 - 3x_2 + 6x_3$$

$$\text{Subject to} \quad 3x_1 - 4x_2 - 6x_3 \leqslant 2$$

$$2x_1 + x_2 + 2x_3 \geqslant 11$$

$$x_1 + 3x_2 - 2x_3 \leqslant 5$$

$$x_1, \quad x_2, \quad x_3 \geqslant 0$$

(a) Set up this problem as a linear programming problem in tableau form.

(b)   Solve this problem.

(c)   How many possible corners are there to the solution space of this problem? How could one determine the coordinates of each of these corners without graphically solving the problem? How does the Simplex Method avoid evaluating Z at each corner of the solution space?

(d)   Distinguish between and discuss the use of artificial and slack variables in the solution of linear programming problems.

13.   A person wishes to select a diet consisting of bread, butter and/or milk, which has a minimum cost, but yields an adequate amount of vitamins A and B. The minimum vitamin requirements are 11 units of A and 10 units of B. The diet should not contain more than 13 units of A, because more may be harmful. Furthermore, he likes milk, and requires that the diet include at least 3 units of milk, but is indifferent to the amount of butter or bread. A dietician has measured the vitamin contents and found, per unit of product, 1 unit of A and 3 of B in bread, 4 units of A and 1 of B in butter, 2 units of A and 2 of B in milk. The market price of one unit of bread, butter, and milk is $2\cancel{c}$, $9\cancel{c}$ and $1\cancel{c}$, respectively. Formulate the problem in primal form. Set up initial Simplex tableaus, and then solve this problem.

14.   Consider this L.P. problem:

$$\text{Maximize } P = 100x_1 + 200x_2 + 50x_3$$

$$\text{Subject to} \quad 5x_1 + 5x_2 + 10x_3 \leqslant 1000$$

$$10x_1 + 8x_2 + 5x_3 \leqslant 2000$$

$$10x_1 + 5x_2 \quad\quad \leqslant 500$$

$$x_1, x_2, x_3 \geqslant 0$$

This L.P. formulation corresponded to a situation in which a company had 1000 tons of ore b_1, 2000 tons of ore b_2 and 500 tons of ore b_3.

Products x_1, x_2 and x_3 can be extracted and blended from these ores. The company wishes to determine how much of each product to make from the available ores to maximize the profit from the overall operation. The requirements on the ores are the following:

Product x_1 requires 5 tons of b_1, 10 tons of b_2 and 10 tons of b_3 per ton.

Product x_2 requires 5 tons of b_1, 8 tons of b_2 and 5 tons of b_3 per ton.

Product x_3 requires 10 tons of b_1, 5 tons of b_2 and none of b_3 per ton.

The manufacturer will make $100 profit on each ton of product x_1, $200 per ton on x_2 and $50 per ton on x_3. After solving this problem, the final tableau is shown below:

FINAL TABLEAU

$c_{(i)}$	BASIS	V_1	V_2	V_3	V_4	V_5	V_6	b_i
50	x_3	−0.5	0	1	0.1	0	−0.1	50
0	x_5	−3.5	0	0	−0.5	1	−1.1	950
200	x_2	2	1	0	0	0	0.2	100
	c_j	100	200	50	0	0	0	———
	ΔOF_j	−275	0	0	−5	0	−35	22,500

(a) Identify values for the following columns in the final tableau

   (i)   The solution (or requirements) vector

   (ii) The slack vectors

   (iii) The structural vectors (for each decision variable)

(b) Give the optimal solution for the problem in terms of the particular situation described above; i.e. in terms of the products being extracted and blended from the available ore. Be complete in your answer.

(c) Write the dual of the initial L.P. problem.

(d) Either in general terms or in terms of the above problem, give a physical interpretation for

   (i)   the "b_i/a_{ij} rule" (choice of outgoing vector)

   (ii) the Simplex criterion (choice of incoming vector)

(e) In the final Simplex tableau of the primal problem identify the *basic* variables and the *nonbasic* variables.

(f) If the profit coefficient on x_1 were to change from 100 to 400, perform sensitivity analysis to determine if the optimality of the final tableau is destroyed.

(g) If, because of the change in (f), the tableau is no longer optimal, find the new optimal solution.

(h)   Suppose that $\Delta OF_4 = 0$ in the final tableau. What would this have meant relative to ΔOF_2, ΔOF_3 and ΔOF_5?

(i)   If two values of b_i/a_{ij} had been equal, nonzero and finite, how would the tie have been broken?

(j)   How could you tell when a problem such as the above (maximization) was unbounded from above?

15.   The attack power of an aircraft command is evaluated as the sum of three factors:

$$A.P. = Speed + Armament + Range$$

A certain command consists of the following items:

Aircraft Type	Attack Power Rating				Crew Size	No. of Planes
	Speed	Armament	Range	Total		
Fighter	4	2	1	7	2	36
Bomber	2	4	4	10	8	12

The total number of men available for making up crews is 120. In a particular mission it is required that the number of fighters not be less than the number of bombers and that the total *speed* rating of the mission be at least 60.

What is the proper assignment of aircraft that will yield the maximum attack power for this mission? Assume the appropriate mix of personnel is available.

16.   A firm manufactures three types of electrical devices, type R9, type S2 and type M18 on which it makes a gross profit of $0.40, $0.30 and $0.45 per device, respectively, Each type R9 device requires twice as much machine time as a type S2 device and 80 percent as much machine time as a type M18 device. If only type R9 devices were manufactured, there would be enough machine time per day to make exactly 1200 of them. The supply of copper wire is sufficient for only 1000 devices to be made each day (all types combined). The type S2 device requires a ceramic insulator of which only 700 are available per day, and the high-skilled labor needed to gauge type M18 devices

permits no more than 400 of these devices to be produced each day. The problem you have been given is to maximize gross profit subject to all the physical limitations stated above.

17. A furniture manufacturer wants to determine how many tables, chairs, desks and/or bookcases to make to optimize utilization of his available resources. These products use two different types of lumber, and he has on-hand 1500 board-feet of the first type and 1000 board-feet of the second type. He has 800 man-hours of labor available for the total job. His sales forecast plus his backorders require him to make at least 40 tables, 130 chairs, 30 desks and no more than 10 bookcases. Each table, chair, desk and bookcase requires 5, 1, 9 and 12 board-feet, respectively, of the first type of lumber; and 2, 3, 4 and 1 board-feet, respectively, of the second type. A table requires 3 man-hours to make, a chair requires 2 man-hours, a desk 5 man-hours and a bookcase 10 man-hours. The manufacturer makes a total profit of $12 on a table, $5 on a chair, $15 on a desk and $10 on a bookcase. Formulate a linear programming problem such that profits are maximized and solve it with the Simplex Method.

18. Formulate the dual of the linear programming problem included as Example 2 in the text (page 124). Add the necessary slack and/or artificial variables and solve the dual problem by using the Simplex Method. Give an economic interpretation of the dual decision variables.

19. Find the optimal solution to the dual problem formulated on pp. 131-32 and compare with the optimal solution to the corresponding primal problem.

SPECIAL FORMS OF LINEAR PROGRAMMING

INTRODUCTION

There are special classes of linear programming problems that are most easily solved by specialized solution procedures, or algorithms. Although these problems can be solved by the Simplex Method, the algorithms presented in this chapter may be used to obtain a solution with far less effort. The names of these special classes of problems may be misleading. The transportation algorithm can be used to divide job lots and assign them to machines to minimize production costs, and the assignment algorithm may be used in a machine replacement problem. A general discussion of each class of problem and several examples of each will enable the reader to recognize many real-world problems that can be optimized with these algorithms.

THE
TRANSPORTATION
PROBLEM

a general discussion

In the classical transportation problem, factories supply a particular type of product to one or more warehouses. There is a cost of transporting the product from each factory to each warehouse. An illustration of these costs in matrix form is given in Table 6-1. The demand at each warehouse and the supply at each factory are also given in Table 6-1. It can be seen that total demand equals total supply. This is one of the basic assumptions that underlies the transportation algorithm. In many realistic problems this assumption is not satisfied. A simple procedure for overcoming this obstacle, which involves the creation of dummy factories or dummy warehouses, will be presented later.

TABLE 6-1   MATRIX OF TRANSPORTATION COSTS (¢/UNIT)

Factory	El Paso	Tiburon	Boulder	Tulsa	Supply at factories (1,000 units)
Denver	10	5	1	5	10
Houston	2	5	7	6	15
Los Angeles	12	2	6	13	25
Demand at warehouses (1,000 units)	8	20	12	10	50 Total shipments

TABLE 6-2   ONE FEASIBLE SOLUTION TO THE TRANSPORTATION PROBLEM

Factory	El Paso	Tiburon	Boulder	Tulsa	Supply (1,000's)
Denver	8 — 10	2 — 5	1	5	10
Houston	2	15 — 5	7	6	15
Los Angeles	12	3 — 2	12 — 6	10 — 13	25
Demand (1,000's)	8	20	12	10	

(Thousands of units shipped over designated route)

One feasible solution for the transportation problem above is given in Table 6-2. The reader can easily verify that all supply and demand constraints are satisfied. The cost of this solution is $3,730 (i.e. $.10 × 8,000 + $0.05 × 2,000 + $.05 × 15,000 + $.02 × 3,000 + $.06 × 12,000 + $.13 × 10,000 = $3,730), which is obtained by multiplying the transportation cost per unit times the number of units shipped over a particular route.

An optimum solution is one that satisfies all supply and demand constraints with non-negative shipments, and gives the lowest possible cost. (There may be multiple optimum solutions that give the same minimum cost.)

There are many real-world situations that can be formulated and solved by using the transportation algorithm. For example, a number of empty railroad cars may be located at certain cities, which can be thought of as "factories." These cars may be required at other locations, which are analogous to warehouses. In this situation the objective is to redistribute railroad cars at a minimum transportation cost. As another example, several lots of jobs may arrive at a lathe department. The cost of finishing a part will depend on the type of lathe to which it is assigned. Job lots may be assigned to different lathes. Any lot may be broken down with its units assigned to different lathes. In this example, the job lots may be viewed as factories; the lathes as warehouses. The objective is to minimize the total cost of finishing the job lots.

linear programming formulation

The information of Table 6-2 is shown in terms of its general notation in Table 6-3. The amount shipped from factory i to warehouse j is x_{ij}. The cost of shipping x_{ij} units from factory i to warehouse j is $c_{ij}x_{ij}$. Here c_{ij} is the cost of shipping a single unit of merchandise from a factory to a warehouse. The objective is to determine a shipping schedule that will minimize the total shipping charges, TC, where

$$TC = \sum_{i=1}^{m} \sum_{j=1}^{n} c_{ij}\, x_{ij}$$

and there are m factories and n warehouses.

The factory capacity at factory i is a_i. One constraint on our solution is that all production must be shipped

$$\sum_{j=1}^{n} x_{ij} = a_i \text{ for factories } i = 1,2,\text{- - -},m$$

A second constraint is that all warehouse requirements must be met (demand at warehouse j is b_j):

$$\sum_{i=1}^{m} x_{ij} = b_j \text{ for warehouses } j = 1,2,\text{- - -},n$$

TABLE 6-3
THE LINEAR PROGRAMMING FORMULATION
OF THE TRANSPORTATION PROBLEM

Warehouse

Factory	1	2	3	4	Supply
1	c_{11} x_{11}	c_{12} x_{12}	c_{13} x_{13}	c_{14} x_{14}	a_1
2	c_{21} x_{21}	c_{22} x_{22}	c_{23} x_{23}	c_{24} x_{24}	a_2
3	c_{31} x_{31}	c_{32} x_{32}	c_{33} x_{33}	c_{34} x_{34}	a_3
Demand	b_1	b_2	b_3	b_4	

This will result in the following:

$$\sum_{i=1}^{m} \sum_{j=1}^{m} x_{ij} = \sum_{i=1}^{m} a_i = \sum_{j=1}^{n} b_j$$

All shipments must be positive or zero, so

$$x_{ij} \geqslant 0 \text{ for } i = 1,2,- - ,m$$
$$j = 1,2,- - ,n$$

This will result in a total of:

(a) m constraints on supply

(b) n constraints on demand

(c) n × m non-negativity constraints.

Since supply must equal demand, any one of the m supply or n demand constraints can be derived from the remaining constraints. Therefore, there are m + n − 1 linearly independent constraints. From the previous discussion of linear programming, a basic solution for the transportation problem is one that has m + n − 1 variables with nonzero values. If there are less than m + n − 1 variables with nonzero values in a solution, that solution is degenerate. Although the transportation algorithm is designed to manipulate basic feasible solutions, there is a simple procedure for treating degenerate solutions, which will be discussed later.

In some problems the variables may be continuous, such as barrels of oil or pounds of grain. However, the units can also be discrete, such as automobiles or computers. Conventional linear programming will give only an approximate solution if the variables are discrete. The transportation algorithm may be used for both continuous and discrete problems.

the transportation algorithm

The first step of the algorithm is to obtain an initial basic feasible solution. The initial solution is checked to determine if it is optimal. If not, shipments are re-assigned to give an improved solution. This procedure of checking for optimality and reassigning shipments is repeated until an optimum solution is located.

Obtaining a Feasible Solution.   One means of obtaining a feasible solution is to satisfy the first warehouse from the first factory and then assign any remaining production from the first factory to the second, third, etc., warehouses successively until it is depleted. The second factory's production is assigned in a similar manner, beginning with the first warehouse having unsatisfied demand. A solution obtained in this manner was shown in Table 6-2. Because this method begins in the upper left hand corner of the table, it is referred to as the "Northwest Corner Rule."

The Northwest Corner Rule will give a feasible solution, but one that is often not optimal. There is a procedure for obtaining an initial solution that is frequently the optimum solution. A modification of this procedure, Vogel's Approximation Method, is presented here. Often the initial feasible solution resulting from application of Vogel's Method will produce the optimal assignment of sources to destinations.

In the transportation problem, the incremental cost will be examined. In Table 6-1, it can be seen that the production at Houston has to be shipped to one or more of the four locations. The cheapest possibility, from Houston's standpoint, is to ship as much as possible to El Paso, with the remainder going to Tiburon. If any is shipped to Boulder it will result in an incremental cost of 7 − 2 = 5 over the cheapest alternative, El Paso. By subtracting the smallest cost in each row from every other cost in that row, the matrix of Table 6-4 is derived from that of Table 6-1. Every warehouse will also have to be supplied, so the lowest cost in each column of the matrix of Table 6-4 should be subtracted from every other cost in its respective column. In Table 6-4, every column contains a zero except the Tulsa column. By subtracting zero from the first three columns and four from each cost in the Tulsa column, the matrix of Table 6-5 is obtained.

A solution that minimizes these incremental costs will minimize the total costs. The transportation algorithm can be applied directly to the total cost matrix given in Table 6-1 or to this matrix of incremental costs as shown in Table 6-5. Incremental costs are used in the following discussion since:

(a) Incremental costs will simplify calculations in many problems.

(b) Incremental costs make more apparent the relative attractiveness of different shipping possibilities.

(c) Incremental costs should make it easier for the reader to fully understand the transportation algorithm.

In Table 6-5, the difference between the smallest cost and the second smallest cost in each row and each column is shown in the margin as row penalties and column penalties. If the Los Angeles factory cannot ship its production to Tiburon, then the next cheapest alternative is to ship to Boulder, which results in an increase of four units in the cost per item. This is a penalty that must be paid if production cannot be shipped along the least expensive route. Similarly, there is a penalty of nine that must be paid if the El Paso demand is not satisfied by a shipment from the Houston factory.

TABLE 6-4   TRANSPORTATION MATRIX
INCREMENTAL FACTORY COSTS

		Warehouse		
Factory	El Paso	Tiburon	Boulder	Tulsa
Denver	9	4	0	4
Houston	0	3	5	4
Los Angeles	10	0	4	11

The largest penalty is nine. To avoid paying this, we must ship as much as possible from Houston to El Paso. This will be eight units, since eight units will satisfy El Paso's demand. This shipment is recorded in Table 6-6a and the Houston supply and El Paso demand are reduced by eight units. An asterisk is placed in every remaining block in the El Paso column to indicate that these squares are no longer eligible for assignment. New row and column penalties are calculated, just as those of Table 6-5, except that squares with shipments and asterisks are ignored. For example, the least expensive location that Houston can ship to is Tulsa with an incremental cost of 0; the second cheapest is Tiburon,

TABLE 6-5   INCREMENTAL COSTS

Factory	El Paso	Tiburon	Boulder	Tulsa	Row Penalty
	Warehouse				
Denver	9	4	0	0	0
Houston	0	3	5	0	0
Los Angeles	10	0	4	7	4
Column Penalty	9	3	4	0	

with an incremental cost of 3. The new row penalty is three. The column penalty is omitted since El Paso's demand is satisfied. For clarity, the revised information is transferred to Table 6-6b.

The procedure discussed above is applied to Table 6-6b. The Los Angeles factory and the Boulder warehouse both have penalities of four. In case of a tie an arbitrary decision may be made. In this second step, ten units are shipped from Denver to Boulder. In the third step of Table 6-6c, seven units are shipped from Houston to Tulsa.

At this stage, it is apparent that the Los Angeles factory must ship twenty units to Tiburon, two units to Boulder and three units to Tulsa. This initial feasible solution is shown in Table 6-6d. If these shipment amounts are multiplied by their respective c_{ij}, the cost of this solution is found to be \$1,590 (\$.01 × 10,000 + \$.02 × 8000 + \$.06 × 7000 + \$.02 × 20,000 + \$.06 × 2000 + \$.13 × 3000 = \$1,590), which improves considerably upon the Northwest Corner Rule solution of \$3,730.

Improving a Feasible Solution.   To improve a feasible solution, two vectors r_i, i=1,2,- - - ,m and s_j, j=1,2,- - -,n will be developed as shown in Table 6-7. In developing r_i and s_j, only squares to which allocations have been made will be considered. The objective is to assign r_i and s_j values such that the c_{ij} of every square with an allocation is equal to its associated $r_i + s_j$. The reader can verify from Table 6-7(a) that, for every square with an allocation, $c_{ij} = r_i + s_j$. The r_i and s_j values are determined in sequence as indicated by the numbers in parentheses in Table 6-7(a). We arbitrarily chose the Denver to Boulder square as a starting point. Since the $c_{ij} = c_{13} = 0$, we may set $r_1 = s_3 = 0$, then $c_{13} = r_1 + s_3 = 0$. This is the first step as indicated by (1) in Table 6-7(a). Notice that $c_{3,3} = 4$, and $s_3 = 0$. Since $c_{33} = r_3 + s_3$, $4 = r_3 + 0$, and as a second step (2), we must set $r_3 = 4$. We proceed in this manner until we have determined all values of r_i and

Special Forms of Linear Programming

TABLE 6-6   CALCULATION OF PENALTIES AND ASSIGNMENTS

(a) First step

Warehouse

Factory	El Paso	Tiburon	Boulder	Tulsa	Row Penalty	Supply
Denver	* 9	4	0	0	0	10
Houston	8 0	3	5	0	Ø3	1̶5̶ 7
Los Angeles	* 10	0	4	7	4	25
Column Penalty	9̶ –	3	4	0		
Demand	8̶ 0	20	12	10		

(b) Second step

Warehouse

Factory	El Paso	Tiburon	Boulder	Tulsa	Row Penalty	Supply
Denver	* 9	* 4	10 0	* 0	Ø–	1̶0̶ 0
Houston	8 0	3	5	0	3	7
Los Angeles	* 10	0	4	7	4	25
Column Penalty	–	3	4̶1	Ø 7		
Demand	0	20	1̶2̶ 2	10		

s_j. As a third step (3), $c_{32} = r_3 + s_2$, $0 = 4 + s_2$, we must set $s_2 = -4$. If the student begins in a different square (e.g. Los Angeles to Tulsa), he may get a different set of r_i and s_j values. These values, however, should lead to an improved solution.

If each r_i were subtracted from every incremental cost in its associated row, and each s_j were subtracted from every incremental cost in its associated column, the result would be a new matrix of incremental costs. In this new matrix the incremental cost of every square with an assignment would be zero. The incremental cost of every other square is $c_{ij} - (r_i + s_j)$. These are shown as circled numbers in Table 6-7(b). Only one of these, $c_{14} - (r_1 + s_4) = 0 - (0 + 3) = -3$, the Denver to Tulsa shipment, is negative. If shipments can be rearranged to make an assignment in this square, the incremental cost will be decreased by three for each unit shipped in this square.

(c) Third step

Warehouse

Factory	El Paso	Tiburon	Boulder	Tulsa	Row penalty	Supply
Denver	9 *	4 *	0 10	0 *	—	0
Houston	0 8	3 *	5 *	0 7	~~3~~ —	~~70~~ 0
Los Angeles	10 *	0	4	7	4	25
Column penalty	—	3	1	7		
Demand	0	20	2	~~10~~ 3		

(d) Initial feasible solution with Vogel's Approximation Method

Warehouse

Factory	El Paso	Tiburon	Boulder	Tulsa	Supply
Denver	9	4	0 10	0	10
Houston	0 8	3	5	0 7	15
Los Angeles	10	0 20	4 2	7 3	25
Demand	8	20	12	10	

To assign shipments to a vacant square, other shipments must be adjusted so that the supply and demand constraints will not be violated. There is a graphical procedure that greatly simplifies this process. The analyst should begin in the square that has the most negative $c_{ij} - (r_i - s_j)$. There he should place a "+" and then draw a series of horizontal and vertical dashed lines as illustrated in Table 6-8. The procedure for drawing these dashed lines is the following. A line may change direction in a square only if there is an assignment in that square. The first time the line changes direction a "−" should be placed in the square. Then "+'s" and "−'s" should alternate. The dashed line should end in the same square in which it began.

To make a reassignment and still satisfy constraints, shipments are added to the squares with "+'s" and subtracted from those with "−'s". The amount of shipment that can be transferred is limited by the least amount

TABLE 6-7   IMPROVING A FEASIBLE SOLUTION

(a) r_i, s_j vectors

Warehouse

Factory	El Paso	Tiburon	Boulder	Tulsa	r_i
Denver	9	4	0 10	0	0 (1)
Houston	0 8	3	3	0 7	−3 (5)
Los Angeles	10	0 20	4 2	7 3	4 (2)
s_j	3 (6)	−4 (3)	0 (1)	3 (4)	

(b) Calculating penalty numbers

Warehouse

Factory	El Paso	Tiburon	Boulder	Tulsa	r_i
Denver	9 ⑥	4 ⑧	0 ⑩	0 ⊝③	0
Houston	0 8	3 ⑩	3 ⑥	0 ⑦	−3
Los Angeles	10 ③	0 20	4 2	7 3	4
s_j	3	−4	0	3	

assigned to a square with a "−", in this case, the three units shipped from Los Angeles to Tulsa. If more than three units were transferred, a negative shipment would result, which is not permissible. The reassignment of these three units is illustrated in Table 6-9a. This was obtained by adding three units to every square with a "+" and subtracting three units from every square with a "−" in Table 6-8.

The values of r_i and s_j are recomputed for this solution and again the incremental costs of vacant squares are computed in Table 6-9b. Since all of the incremental costs are positive, reassignment of shipments will result in an increased total cost. Therefore, this solution is an optimum solution. If any $[c_{ij} - (r_i + s_j)]$ were negative, the procedure would have been reapplied.

The dashed lines form a rectangle in the example problem of Table 6-8. This is not always the case. If we had to reassign a shipment to the

TABLE 6-8   REASSIGNING SHIPMENTS

Warehouse

Factory	El Paso	Tiburon	Boulder	Tulsa	Supply
Denver	9	4	− ⎯ 0 ⎯ + 10	0	10
Houston	0 / 8	3	3	0 / 7	15
Los Angeles	10	0 / 20	4 / + 2	7 / − 3	25
Demand	8	20	12	10	

Houston-Tiburon square, based on the results of Table 6-9(a), the dashed lines would follow the path illustrated in Table 6-10.

Degeneracy.   It is possible to have a solution with less than n + m − 1 shipments, which is a degenerate solution. This may occur with the initial feasible solution, or during the reassignment of shipments. If we alter the demand at El Paso and Boulder in our original problem and obtain an initial solution as shown in Table 6-11a, we see that there are only five shipments, where there are n + m − 1 = 4 + 3 − 1 = 6 required for a basic solution. In this case, we assign some negligible amount of shipment, ϵ, to a low-cost square (the Denver to Tulsa square). This is illustrated in Table 6-11a. There is no change in the solution procedure. In Table 6-11b, the incremental cost is calculated for vacant squares. In the reassignment, five units are added to the Los Angeles to Boulder and the Denver to Tulsa squares, giving a shipment of 5 + ϵ, which is equivalent to 5 for all practical purposes. The incremental costs of this reassignment are calculated in Table 6-11c. Since all incremental costs are positive, the optimum solution is given in Table 6-11c. Notice that this reassignment resulted in a nondegenerate solution. It is possible, however, for the optimum solution to be degenerate. During the solution, no shipment can be subtracted from a square with an ϵ shipment.

Unequal Supply and Demand.   In many realistic problems the supply at the factories and the demand at the warehouses are not equal. This problem can be easily overcome by creating dummy factories or warehouses. If demand exceeds supply, a dummy factory can be created. The cost of supplying a warehouse from this dummy factory is the per unit cost of the warehouse not receiving this demand. If supply exceeds demand, a dummy warehouse is created. The cost of a factory supplying this dummy warehouse is the per unit cost of not producing the goods.

TABLE 6-9   AN IMPROVED SOLUTION

(a) The new assignment

Warehouse

Factory	El Paso	Tiburon	Boulder	Tulsa
Denver	[9]	[4]	[0] 7	[0] 3
Houston	[0] 8	[3]	[3]	[0] 7
Los Angeles	[10]	[0] 20	[4] 5	[7]

(b) Checking for optimality

Warehouse

Factory	El Paso	Tiburon	Boulder	Tulsa	r_i
Denver	[9] ⑨	[4] ⑧	[0] 7	[0] 3	0
Houston	[0] 8	[3] ⑦	[3] ③	[0] 7	0
Los Angeles	[10] ⑥	[0] 20	[4] 5	[7] ③	4
s_j	0	−4	0	0	

TABLE 6-10   A POSSIBLE REASSIGNMENT PATTERN

Warehouse

Factory	El Paso	Tiburon	Boulder	Tulsa	Supply
Denver	[9]	[4]	[0] − 7	[0] + 3	10
Houston	[0] 8	[3] + *	[3]	[0] 7	15
Los Angeles	[10]	[0] − 20	[4] 5 +	[7]	25
Demand	8	20	12	10	

TABLE 6-11 DEGENERACY

(a) A degenerate initial solution

Warehouse

Factory	El Paso	Tiburon	Boulder	Tulsa	Supply
Denver	9	4	0 10	0 ε	10
Houston	0 10	3	5	0 5	15
Los Angeles	10	0 20	4	7 5	25
Demand	10	20	10	10	

(b) Evaluation of incremental costs

Warehouse

Factory	El Paso	Tiburon	Boulder	Tulsa	r_i
Denver	9 ⑨	4 ⑪	0 − 10	0 + ε	0
Houston	0 10	3 ⑩	5 ⑤	0 5	0
Los Angeles	10 ③	0 20	4 + ③	7 − 5	7
S_j	0	−7	0	0	

(c) Reassignment and evaluation

Warehouse

Factory	El Paso	Tiburon	Boulder	Tulsa	r_i
Denver	9 ⑨	4 ⑧	0 5	0 5	0
Houston	0 10	3 ⑦	5 ⑤	0 5	0
Los Angeles	10 ⑥	0 20	4 5	7 ③	4
S_j	0	−4	0	0	

In Table 6-12a an example is presented that has supply greater than demand. A dummy warehouse is created and a demand of ten units is assigned to this dummy warehouse to balance supply and demand. In most applications when a dummy facility is employed, a cost of zero is associated with the dummy facility. This implies that there is no cost of shipping nonexistent units. In this example, costs are associated with the dummy facility. These represent the economic differences in production at the different factories (i.e. it is more economical to produce less than capacity at the Houston plant than at the Denver or Los Angeles plants). These costs are included in this example to make the reader aware of the realistic economic factors involved in a problem of this type.

From Table 6-12c, it can be seen that an optimum solution involves producing at five units under capacity at both Houston and Los Angeles.   This is because five units appear in the dummy warehouse column for these two factors.

THE
TRANSSHIPMENT
PROBLEM

formulation as a
transportation problem

The transshipment problem is a special case of the transportation problem. In the transportation problem all shipments are made directly from factories to warehouses. In reality one factory may ship to another factory where the shipments are combined and forwarded to a plant. A warehouse may receive shipments, break them up and ship part to another warehouse. The transshipment problem takes these realities into account by permitting any factory or warehouse to receive goods and forward them. This situation is illustrated in Table 6-13, in which the transportation example is now treated as a transshipment problem. This cost matrix reflects the appropriate transshipment costs. Notice that all diagonal elements have a zero cost. This represents the cost of shipping from a location to itself (from Denver to Denver, etc). Shipments assigned to these diagonal elements represent items that are not transshipped; they will be ignored in the final solution.

The limit on the amount that can be transshipped is the total shipments

$$\text{i.e.} \sum_{i=1}^{m} a_i$$

from the basic transportation problem. In the example of Table 6-1, this is

TABLE 6-12   SUPPLY EXCEEDS DEMAND

(a) Statement of problem

Warehouse

Factory	El Paso	Tiburon	Boulder	Tulsa	Dummy	Supply
Denver	10	5	1	5	3	20
Houston	2	5	7	6	2	15
Los Angeles	12	2	6	13	3	25
Demand	8	20	12	10	10	60

(b) Calculation of incremental costs and initial assignment

Warehouse

Factory	El Paso	Tiburon	Boulder	Tulsa	Dummy	Penalty	Supply
Denver	9 / *	4 / *	0 / 12	0 / 8	2 /	Ø 2	2̶0̶ 8̶ 0
Houston	0 / 8	3 / *	5 / *	0 / 2	0 / 5	0	1̶5̶ 7
Los Angeles	10 / *	0 / 20	4 / *	7 / *	1 / 5	1̶ 6	2̶5̶ 5̶ 0
Penalty	Ø—	3̶—	4̶—	0	1̶ 2		
Demand	8̶ / 0	2̶0̶ / 0	1̶2̶ / 0	1̶0̶ / 2	1̶0̶ / 5		

(c) Evaluation of the initial assignment

Warehouse

Factory	El Paso	Tiburon	Boulder	Tulsa	Dummy	r_i
Denver	9 / ⑨	4 / ⑤	0 / 12	0 / 8	2 / ②	0
Houston	0 / 8	3 / ④	5 / ⑤	0 / 2	0 / 5	0
Los Angeles	10 / ⑨	0 / 20	4 / ③	7 / ⑥	1 / 5	1
s_j	0	−1	0	0	0	

179

50,000 units. There is a total of 50,000 units that may be shipped to any factory or warehouse. This is added to the supply and demand columns as shown in Table 6-13. For example, supply at Houston is 65,000 units. There are 15,000 units manufactured at Houston and 50,000 units that may be transshipped from Houston.

The transshipment problem is then solved using the transportation algorithm. The solution is shown in Table 6-14. The solution in Table 6-14 is the same solution derived for the basic transportation problem. There is no benefit in making transshipments, for this particular example.

TABLE 6-13   FORMULATION OF A TRANSSHIPMENT PROBLEM

				Destination				
<ins>Source</ins>	Denver	Houston	Los Angeles	El Paso	Tiburon	Boulder	Tulsa	Supply (1,000 units)
Denver	0	7	6	10	5	1	5	10 + 50
Houston	7	0	4	2	5	7	6	15 + 50
Los Angeles	6	4	0	12	2	6	13	25 + 50
El Paso	10	2	12	0	4	9	8	0 + 50
Tiburon	5	5	2	4	0	5	14	0 + 50
Boulder	1	7	6	9	5	0	5	0 + 50
Tulsa	5	6	13	8	14	5	0	0 + 50
Demand (1,000 units)	0 +50	0 +50	0 +50	8 +50	20 +50	12 +50	10 +50	

TABLE 6-14   SOLUTION TO THE TRANSSHIPMENT PROBLEM

			Warehouse				
Factory	Denver	Houston	Los Angeles	El Paso	Tiburon	Boulder	Tulsa
Denver	[0] 50	[7]	[6]	[10]	[5]	[1] 7	[5] 3
Houston	[7]	[0] 50	[4]	[2] 8	[5]	[7]	[6] 7
Los Angeles	[6]	[4]	[0] 50	[12]	[2] 20	[6] 5	[13]
El Paso	[10]	[2]	[12]	[0] 50	[4]	[9]	[8]
Tiburon	[5]	[5]	[2]	[4]	[0] 50	[5]	[14]
Boulder	[1]	[7]	[6]	[9]	[5]	[0] 50	[5]
Tulsa	[5]	[6]	[13]	[8]	[14]	[5]	[0] 50

THE ASSIGNMENT
PROBLEM

a general discussion

The assignment problem is a special case of the transportation problem. It may be viewed as a transportation problem with a demand of one at each destination and a supply of one at each source, where all shipments must be either zero or one. A shipment of one indicates that a source is assigned to a destination. The solution involves assigning n sources to n destinations to minimize the total assignment cost. The assignment matrix must be a square (n × n) matrix. Special procedures may be used when the number of sources and destinations is not equal. These will be illustrated in example problems. While the assignment problem can be solved by the Simplex Method or the transportation algorithm, we will discuss another special algorithm that requires less effort.

Many realistic problems can be solved with the assignment algorithm. If, for example, there are n workers to be assigned to n jobs such that all jobs should be accomplished at minimum cost to the company, then the workers may be viewed as sources and jobs as destinations. The costs for three workers and three jobs are given by a matrix such as that of Table 6-15. It can be seen that the cost of having worker U perform job 2 is 22 units. In the solution there must be exactly one assignment in each column, and one assignment in each row—there must be exactly one worker for each job.

TABLE 6-15   ASSIGNMENT COST MATRIX

<u>Job</u>

<u>Worker</u>	1	2	3
U	43	22	28
V	40	38	37
W	24	25	36

As a second example, we may use the assignment algorithm to assign n job lots to n lathes. If the job lots can be broken up and one job lot assigned to multiple machines, then the transportation algorithm is applicable. If set-up costs are very high and one job lot must be assigned to only one machine, then the assignment algorithm should be used.

Another typical example is the assignment of n independent new machines to n possible locations where the assignment costs are the material handling costs associated with assigning a machine to a particular location.

linear programming
formulation

The linear programming formulation of the assignment problem is straightforward. Using the example of Table 6-15, our objective function is

$$\text{Minimize } Z = 43x_{U1} + 22x_{U2} + 28x_{U3} + 40x_{V1} + 38x_{V2} + 37x_{V3} + 24x_{W1} + 25x_{W2} + 36x_{W3}$$

where

$$x_{ij} = 1 \text{ if worker i (i = U,V,W) is assigned}$$
$$\text{to job j (j=1,2,3)}$$

and

$$x_{ij} = 0 \text{ if worker i is } not \text{ assigned to job j}$$

To avoid assigning two jobs to one worker or two workers to a single job, a set of constraint equations could be written:

$$x_{U1} + x_{U2} + x_{U3} = 1$$
$$x_{V1} + x_{V2} + x_{V3} = 1$$
$$x_{W1} + x_{W2} + x_{W3} = 1$$

Requires that every worker be assigned to exactly one job.

$$x_{U1} + x_{V1} + x_{W1} = 1$$
$$x_{U2} + x_{V2} + x_{W2} = 1$$
$$x_{U3} + x_{V3} + x_{W3} = 1$$

Requires that each job be completed by one worker only.

In general the assignment problem can be summarized as follows:

Minimize $Z = \sum\limits_{i}^{n} \sum\limits_{j}^{n} c_{ij}x_{ij}$ (c_{ij} = cost of assigning worker i to job j)

Subject to $\sum\limits_{i=1}^{n} x_{ij} = 1$

$\sum\limits_{j=1}^{n} x_{ij} = 1$

$x_{ij} = 1$ or 0

For the problem above, one may be tempted to make assignments by trial-and-error in an effort to discover the minimum cost. Unfortunately, for large problems, this approach would be impractical. Even when n=3 (a small problem), there are six possible assignments for the above problem as enumerated below.

Trial Number	Worker-Job Assignment	Cost
1	U-1, V-2, W-3	117
2	U-1, V-3, W-2	105
3	U-2, V-1, W-3	98
4* (optimum)	U-2, V-3, W-1	83*
5	U-3, V-1, W-2	93
6	U-3, V-2, W-1	90

Thus, the optimal assignment would be expressed as in Table 6-16 in matrix form.

TABLE 6-16   OPTIMAL ASSIGNMENT

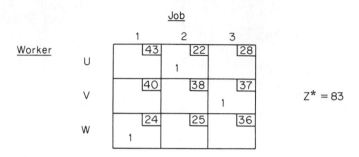

In general there are n! = n(n-1)(n-2)---(3)(2)(1) possible assignments to be considered. Obviously, when n=6 it would be difficult to evaluate all 720 possibilities. To save time, the following systematic solution procedure for this type of minimization problem has been developed.

the assignment
algorithm

The assignment algorithm will be presented first and then two examples will be given to illustrate its application.

Step 1—Subtract the smallest cost in row i from every cost in row i of the matrix.

Step 2—If n zero elements do not result from Step 1, use the transformed matrix of Step 1 to subtract the smallest cost in column j from every cost in column j. In subtracting minimum costs from rows and columns, the principle is that adding or subtracting a constant from all elements in a matrix does not alter the optimum assignment. That is, an optimal assignment for the original cost matrix is also optimum for the transformed matrix.

After completing Step 1, the original cost matrix would be as in Table 6-17 (Step 2 was unnecessary in this problem). The next step involves making the assignments of workers to jobs and checking the solution for optimality.

Step 3—The object now is to make assignments to zero cells such that *all* constraints are met. In our problem, it is possible to make a valid allocation because job 2 can be assigned to worker U, job 3 to worker V and job 1 to worker W (see matrix of Table 6-16). In this case the minimum cost is 22+37+24 = 83.

TABLE 6-17   TRANFORMED COST MATRIX

Job

Worker		1	2	3
U		21 | 1	0	6
V		3	1 | 1	0
W		0 | 1	1	12

Several complications can arise during Step 3. Suppose that the original matrix was modified by changing the cost in cell (W,1) to 38 (it had been 24). The results of Steps 1 and 2 would be as shown in Table 6-18.

TABLE 6-18   TRANSFORMATION OF COST MATRIX

Step 1

Job

	1	2	3
U	21	0	6
V	3	1	0
W	13	0	11

Step 2

Job

	1	2	3
U	18	0	6
V	0	1	0
W	10	0	11

Even though there is at least one zero in every row and column, it is not possible to make an assignment because, for example, worker V may be given jobs 1 or 3. However, there is no other worker who can be assigned job 1 or job 3 at zero incremental cost. Only two feasible, zero-cost assignments can be chosen, so we need to somehow create at least one more cell containing a zero. This is accomplished by a clever but simple technique.

That we do not have an optimum (or even feasible) assignment is apparent by drawing the minimum number of lines necessary to cover all zero elements.

Since less than n = 3 lines can be drawn to cover all zeros, the optimal solution has not yet been determined. If all zeros had been covered by exactly n lines, a necessary condition for an optimum solution would have been satisfied.

To deal with the above situation, find the minimum element *uncovered* by a line. Subtract this number from all uncovered costs in the matrix and add it to every cost at the *intersection* of two lines. Leave all other numbers unchanged. This procedure is the equivalent of subtracting the minimum cost from every

TABLE 6-19   CHECKING FOR AN OPTIMUM SOLUTION

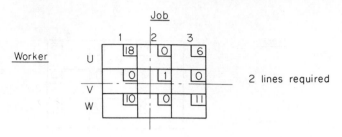

2 lines required

cost in the table, and then adding it to every row and every column with a line through it. In our matrix in Table 6-19 the minimum uncovered element is 6 and the result is as shown in Table 6-20.

TABLE 6-20   AN OPTIMAL SOLUTION

Job

Worker	1	2	3
U	12	* 0 1	0
V	0 1	7	* 0
W	4 1	0	5

In making our assignment, notice there is only one zero in row W and one zero in column 1. It is best to begin making assignments in rows and columns with only one zero. Therefore, let us (arbitrarily) assign worker W to job 2. This *eliminates* the possibility of assigning worker U to job 2 so we place an asterisk in cell (U,2). Then it is only possible to assign worker U to job 3. After assigning worker U to job 3, the possibility of assigning worker V to job 3 is eliminated and an asterisk is placed in this square. Finally worker V is placed on job 1. This allocation of workers to jobs has a total cost of 93 (refer to total costs of Table 6-15). The reader can verify that 93 is the minimum cost by simply evaluating all possible assignments as we did earlier.

It should be noted that if the above procedure for creating additional zero cells had not produced an optimal assignment on the first try, it could have been repeated until an optimal assignment was found. Again, optimality results when (1) the minimum number of lines required to cover all zero cells equals n and (2) the constraints are satisfied.

Example 1:

To further demonstrate the assignment method of linear programming, consider a ship unloading-loading problem. Suppose the value of a ship working-day is estimated to average $20,000. The cost of labor per day to unload and load the ship is negligible in comparison to the value of a ship-working day, so it is desirable to minimize the time that each ship must spend in port. There are four berths for unloading-loading cargos and three ships are due to arrive at the facility when it is empty. Because the type of cargo varies from one ship to another, the productivity of labor at each berth also varies according to which ship is assigned to a particular berth. These estimates of production in terms of tons per hour are shown in Table 6-21(a) for the three incoming ships.

TABLE 6-21(a)   LABOR PRODUCTIVITY (TONS/HR)

		Ship 1	Ship 2	Ship 3
		1	2	3
Berth	1	10	13	16
	2	13	10	14
	3	11	15	18
	4	15	10	8

The tonnage of each ship is: Ship 1 = 2500 tons, Ship 2 = 2300 tons and Ship 3 = 1950 tons. If labor gangs work three shifts each day, the total number of days required to *unload and reload* the three ships in each berth is summarized in Table 6-21(b) (e.g. 5000 tons of cargo would be handled in unloading and reloading Ship 1; for Ship 1 in Berth 1,

$$\frac{5000 \text{ tons}}{10 \text{ tons/hr.} \times 24 \text{ hr./day}} = 20.8 \text{ days}$$

so 20.8 days are required for loading and unloading).

The assignment method must be applied to a square matrix (n × n) of penalty values (or rewards) so we will add a "dummy" ship whose unloading time is zero days. In the final solution the dummy ship will be assigned to one of the four berths. This of course tells us which berth is to remain idle. After

TABLE 6-21(b)   DAYS TO LOAD
AND UNLOAD SHIPS

Ship

		1	2	3
	1	20.8	14.7	8.6
Berth	2	16.0	19.2	11.6
	3	18.9	12.8	9.0
	4	13.9	19.2	20.2

adding the extra column and subtracting the minimum penalty from all cells in each column to find minimum delay assignments, we would have the matrix of Table 6-21(c).

TABLE 6-21(c)   AN OPTIMUM SOLUTION

Ship

		1	2	3	4
	1	6.9	1.9	0 [1]	0
Berth	2	2.1	6.4	3.0	0 [1]
	3	5.0	0 [1]	0.4	0
	4	0 [1]	6.4	11.6	0

We can minimize unloading-loading time by assigning Ship 3 to Berth 1, Ship 2 to Berth 3 and Ship 1 to Berth 4. In this case Berth 2 would not be used. The minimum unloading-reloading *workload* is 13.9 + 12.8 + 8.6 = 35.3 days of effort. (This should not be confused with calendar time.)

As we saw earlier, after Steps 1 and 2 we are sometimes not able to get n or more zero-valued cells in our cost matrix or to make assignments satisfying all constraints on the problem. In Step 3 we discussed a simple technique for dealing with this situation. Example 2, which follows, is intended to illustrate this technique and to present the reader with an extension of a typical man-machine assignment problem.

Example 2:

Suppose five workers are available to work with five existing machines. Machine MA is a new piece of equipment designed to *replace* one of the five older machines. It is not known which of the older machines MA should replace. The costs of assigning each worker to the various machines are given in the matrix of Table 6-22(a), where an "R" means that the indicated man-machine combination is restricted (not possible) for some reason.

TABLE 6-22(a)   ASSIGNMENT COSTS

Worker	M1	M2	M3	M4	M5	MA
A	12	3	6	R	5	9
B	4	11	R	5	R	3
C	8	2	10	9	7	5
D	R	7	8	6	12	10
E	5	8	9	4	6	R

The problem now is to determine the least-cost combination of workers and machines, taking account of the requirement that MA must replace one of the older machines in the optimal assignment. We also would like to know the total cost of this assignment.

It is immediately apparent that a "dummy" worker, F, must be added to make our cost matrix square (i.e. 6 × 6). We will give the dummy worker a cost of 0 on each job so that he will be assigned to exactly one machine (not MA) in the final assignment. The machine to which the dummy worker is assigned will then identify for us the old machine to be replaced by MA.

Because we cannot consider man-machine combinations where R's appear in the matrix in Table 6-22(a), we need to make certain they will not be included in the final assignment. To do this, we will place a very high cost in each element of the matrix where an R appears. For simplicity, use an "M" to denote these high costs (let M = $1000 if you prefer). After adding the dummy worker and inserting M's where needed, the matrix appears as shown in Table 6-22(b). After subtracting the smallest cost in each row from all costs in the row, we obtain Table 6-22(c).

TABLE 6-22(b)   ADDITION OF A DUMMY WORKER

Machine

Worker	M1	M2	M3	M4	M5	MA
A	12	3	6	M	5	9
B	4	11	M	5	M	3
C	8	2	10	9	7	5
D	M	7	8	6	12	10
E	5	8	9	4	6	M
F	0	0	0	0	0	0

TABLE 6-22(c)   INCREMENTAL COSTS

Machine

Worker	M 1	M 2	M 3	M 4	M 5	M A
A	9	0 _1_	3	M−3	2	6
B	1	8	M−3	2	M−3	0 _1_
C	6	* 0	8	7	5	3
D	M−6	1	2	0 _1_	6	4
E	1	4	5	* 0	2	M−4
F	0 _1_	* 0	* 0	* 0	* 0	* 0

Because we must assign MA first (not to the dummy worker), the resultant assignment is indicated in Table 6-22(c). No more than four assignments can be made, so we must resort to the "line covering" procedure described earlier. (Six assignments must be made.)

After covering all zero cells with the least possible number of lines, we would have Table 6-22(d).

Next subtract the least-cost uncovered cell (i.e. 1) from all uncovered cells and add this number to cells where two lines cross (leave all other cells unaltered). This is shown in Table 6-22(e).

TABLE 6-22(d) FIRST APPLICATION OF LINE COVERING PROCEDURE

Machine

Worker	M1	M2	M3	M4	M5	MA
A	9	0	3	M-3	2	6
B	1	8	M-3	2	M-3	0
C	6	0	8	7	5	3
D	M-6	1	2	0	6	4
E	1	4	5	0	2	M-4
F	0	0	0	0	0	0

TABLE 6-22(e) CHECKING FOR AN OPTIMAL ASSIGNMENT

Machine

Worker	M1	M2	M3	M4	M5	MA
A	8	0 (1)	2	M-3	1	6
B	0 (*)	8	M-4	2	M-4 (1)	0
C	5	0 (*)	7	7	4	3
D	M-7	1	1	0 (1)	5	4
E	0 (1)	4	4	0 (*)	1	M-4
F	0 (*)	1	0 (*)	1	0 (1)	1

Now we make assignments of workers to machines in Table 6-22(e), and discover that only five combinations are possible. It is necessary to repeat the line covering procedure until the optimal assignment is determined. These operations are shown in Tables 6-22(f) and 6-22(g).

TABLE 6-22(f) SECOND APPLICATION OF LINE COVERING PROCEDURE

Machine

Worker	M 1	M 2	M 3	M 4	M 5	MA
A	8	0	2	M-3	1	6
B	0	8	M-4	2	M-4	0
C	5	0	7	7	4	3
D	M-7	1	1	0	5	4
E	0	4	4	0	1	M-4
F	0	1	0	1	0	1

TABLE 6-21(g) AN OPTIMAL ASSIGNMENT

Machine

Worker	M1	M2	M3	M4	M5	MA
A	8	0 (*)	1	M-3	0 (1)	6
B	0 (*)	8	M-5	2	M-5 (1)	0
C	5	0 (1)	6	7	3	3
D	M-7	1	0 (*)	0 (1)	4	4
E	0 (1)	4	3	0 (*)	0 (*)	M-4
F	1	2	0 (1)	2	0 (*)	2

Finally, an assignment can be made that satisfies the row-column constraints. We also observe that all zero cells can be covered by no fewer than six lines, so our optimal solution has been determined. The final man-machine combination would be:

Worker A to Machine 5

Worker B to Machine MA (new machine)

Worker C to Machine 2

Worker D to Machine 4

Worker E to Machine 1

Worker F (dummy) to Machine 3

Thus, we would replace Machine 3 by the new machine in addition to making the above noted man-machine assignments. The minimum cost of this set of assignments is:

$$Z = 5 + 3 + 2 + 6 + 5 = 21 \text{ cost units}$$

INTEGER

PROGRAMMING

introduction

Integer programming is required when decision variables must assume integer values. For example, in deciding how many aircraft of different types to produce, it makes no sense to express the optimal production mix in other than integer quantities. An integer linear programming problem is illustrated in Figure 6-1. It can be seen that there is only a limited, discrete number of feasible combinations of decision variable values. The constraints and the objective function are continuous linear equations. If the example of Figure 6-1 is treated as a conventional linear programming problem, the optimum is found to be $X_0 = (1.74, 1.96)$. If we round this solution off to the nearest integer values, we get $X_1 = (2,2)$ which is an infeasible point. The two feasible solutions which are nearest X_0 are $X_2 = (1,2)$ and $X_3 = (2,1)$. However, neither of these points is the optimum solution to our integer linear programming problem. It turns out that the optimum is $X^* = (5,0)$.

Gomory's cutting plane technique is employed in this section to solve the integer linear programming problem. The initial step is to obtain a solution to the continuous problem by using the Simplex Method. If the optimum decision variables have integer values, then this is also the solution to the integer problem. If any of the optimum decision variables have noninteger values, an additional constraint, or cutting plane, is added to the last tableau of the Simplex Method. This cutting plane is formulated so that it reduces the size of the feasible region, but does not eliminate any feasible decision variable values. These cutting planes are added one at a time until an optimal integer solution is reached.

Gomory's cutting plane technique

Gomory's cutting plane technique will be presented through the following.

Example:

An excursion company is considering adding small boats to their fleet. The company has $200,000 to invest in this venture. At present there is an estimated maximum demand of 6,000 customers per season for these tours. The company does not wish to provide capacity in excess of the estimated maximum demand. The basic data are given below for the two types of available boats. The company will make an estimated seasonal profit of $4,000 for each boat of Type 1 and $7,000 for each boat of Type 2. How many boats of each type would you use to maximize profit?

	Type 1	*Type 2*
capacity, $\dfrac{\text{customers}}{\text{season}}$	1,200	2,000
initial cost, $\dfrac{\$}{\text{boat}}$	25,000	80,000

The linear programming problem (illustrated in Figure 6-1) is:

$$\text{Maximize} \quad 4{,}000\,x_1 + 7{,}000\,x_2$$

$$\text{Subject to} \quad 1{,}200\,x_1 + 2{,}000\,x_2 \leqslant 6{,}000$$

$$25{,}000\,x_1 + 80{,}000\,x_2 \leqslant 200{,}000$$

$$x_1, x_2 \geqslant 0 \,; x_1 \; x_2 \text{ are integers}$$

Constraints in the original problem formulation should be transformed so that all coefficients are integers. This is done to facilitate solution of the integer programming problem. No transformation is required in this example. If there were a constraint such as $(3/4)x_1 + (6/4)x_2 \leqslant 48/10$, both sides must be multiplied by 20 so that it becomes $15x_1 + 30x_2 \leqslant 96$. To simplify notation, let us divide the first constraint by 100 and the objective function and second constraint by 1000 and construct the initial tableau.

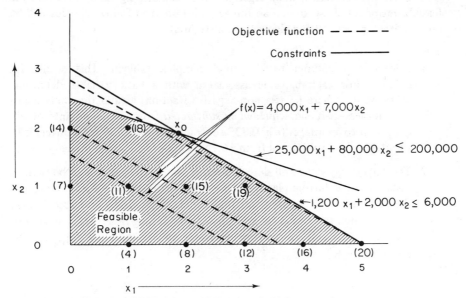

FIGURE 6-1 An Integer Linear Programming Example

c_i	BASIS	V_1	V_2	V_3	V_4	b_i
0	x_3	12	20	1	0	60
0	x_4	25	80	0	1	200

The final tableau in the Simplex solution is:

c_i	BASIS	V_1	V_2	V_3	V_4	b_i	f_{i0}
4	x_1	1	0	0.1739	−0.0435	1.739	0.739
7	x_2	0	1	−0.0543	0.0261	1.956	0.956
	ΔOF_j	0	0	−0.3155	−0.0087	20.65	

Since the solution is noninteger, we must add cutting planes to reduce the feasible region until we obtain an integer solution. The following steps will be used to develop new cutting planes (or constraints).

1. Add a new column to the final Simplex tableau. This is the f_{i0} column. For each b_i value associated with a basic variable determine an f_{i0} value, where f_{i0} is a non-negative fraction, greater than or equal to zero but less than one, which when *subtracted* from a given noninteger will convert it to an integer (e.g. 0.739 subtracted from 1.739 will convert it to an integer; 0.25 subtracted from -6.75 will convert it to an integer).

2. The largest f_{i0} value will determine the row of the tableau to be used in constructing a cutting plane. In the above tableau $f_{20} = 0.956$ designates the second row to be used for this purpose since $0.956 > 0.739$ (i.e. $f_{20} > f_{10}$). When ties occur, an arbitrary choice among tied rows is made. For each a_{ij} coefficient in this row determine an f_{ij} value, just as f_{i0} was determined for b_i.

	x_1	x_2	x_3	x_4	b_i
Row 2	0	1	-0.0543	0.0261	1.956
f_{2j} values	0	0	0.9457	0.0261	0.956
Integer value	0	1	-1	0	1

The f_{2j} values give a new constraint

$$0x_1 + 0x_2 + 0.9457x_3 + 0.0261x_4 \geqslant 0.956$$

Adding a surplus variable x_5 gives:

$$0x_1 + 0x_2 + 0.9457x_3 + 0.0261x_4 - x_5 = 0.956$$

(The logic that permits construction of this new "$\geqslant$" constraint will be presented at the completion of the example.)

3. The new constraint is added to the final Simplex tableau. The *incoming* variable is the one that will cause the smallest decrease in the objective function as indicated by the ΔOF_j values of the final Simplex tableau. An alternative rule (sometimes more efficient) is to select the incoming variable as that having the maximum quotient of $\Delta OF_j/a_{ij}$ for nonbasic variable j, where $a_{ij} < 0$. Our incoming variable will be x_4 with a ΔOF_4 of -0.0087. The *outgoing* variable, x_5, is always that associated with the constraint just annexed (in this case, row 3).

c_i	BASIS	V_1	V_2	V_3	V_4	V_5	b_i
4	x_1	1	0	0.1739	-0.0435	0	1.739
7	x_2	0	1	-0.0543	0.0261	0	1.956
0	x_5	0	0	0.9457	0.0261	-1	0.956 $\longrightarrow$

Applying the Simplex procedure, we get the following tableau:

c_i	BASIS	V_1	V_2	V_3	V_4	V_5	b_i	f_{i0}
4	x_1	1	0	1.750	0	- 1.667	3.333	.332
7	x_2	0	1	-1	0	1	1.000	.000
0	x_4	0	0	36.234	¹	-38.3142	-0.333	–
	ΔOF_j	0	0	0	0	-0.333	20.333	

In our new solution X = (3.333, 1.000). Since x_1 has the maximum f_{i0}, it is used to determine the next cutting plane. Upon adding another column V_6, and another constraint, our next tableau becomes:

c_i	BASIS	V_1	V_2	V_3	V_4	V_5	V_6	b_i
4	x_1	1	0	1.750	0	-1.667	0	3.333
7	x_2	0	1	-1	0	1	0	1.000
0	x_4	0	0	36.234	1	-38.3142	0	36.628
0	x_6	0	0	.750	0	0.333	-1	0.333 $\longrightarrow$

Applying the Simplex procedure we obtain:

c_i	BASIS	V_1	V_2	V_3	V_4	V_5	V_6	b_i
4	x_1	1	0	5.5	0	0	5	5
7	x_2	0	1	- 325	0	0	3	0
0	x_4	0	0	122.44	1	0	-114.94	74.94
0	x_5	0	0	2.25	0	1	-3	1.00
	ΔOF_j	0	0	-2197	0	0	-41	20

The above tableau gives us the optimum integer solution, X* = (5,0).

Cutting Plane Logic. The first cutting plane added is based on row 2 of the final Simplex tableau for noninteger decision variables (see p. 195).

$$0x_1 + x_2 - 0.0543x_3 + 0.0261x_4 = 1.956$$

If we state each coefficient as some integer plus a fraction between zero and one, we get

$$(0 + 0)x_1 + (1.00 + 0)x_2 + (-1.00 + 0.9457)x_3 +$$
$$(0 + 0.0261)x_4 = 1.000 + 0.956$$

If we now move the integer quantities to the right, we get

$$0.9457x_3 + 0.261x_4 = 0.956 + \text{(integer values)}$$

Since all variables must be zero or positive, the expression on the left must be positive, and therefore the expression on the right must be positive also. This gives us an expression that must be satisfied for an integer problem:

$$0.9457x_3 + 0.261x_4 \geqslant 0.956$$

After including a surplus variable we get our first cutting plane:

$$0.9457x_3 + 0.261x_4 - x_5 = 0.956$$

This was the constraint added to the tableau on page 197. The logic used to construct the second cutting plane is identical to that for the first.

Let us quickly consider another integer programming problem:

$$\text{Minimize} \quad x_1$$
$$\text{Subject to} \quad -2x_1 + 12x_2 \geqslant 6$$
$$\frac{2x_1}{3} + x_2 \geqslant 6$$
$$x_2 \leqslant 3$$
$$x_j \geqslant 0 \text{ and } x_j \text{ must be integers}$$

After adding the appropriate slack, surplus and artificial variables and working this problem with the Simplex Method, we find that $x_1 = 4.5$ in the optimal solution. By adding the necessary cutting planes (constraints) to the last tableau of the Simplex procedure, you should be able to show that the optimal integer solution to this problem is $x_1 = 6$; x_2 may assume a value of 2 or 3.

Two articles are included in the Appendix that illustrate several special forms of linear programming. The first is entitled "Linear Programming Without the Math," by Edward Cochran. The other article is "Linear Programming to Aid Resource Allocation in R&D," written by A.G. Lockett and A.E. Gear.

SUGGESTED ADDITIONAL READINGS

Hillier, Fredrick S., and Lieberman, Gerald J. *Introduction to Operations Research.* San Francisco: Holden-Day, Inc., 1969.

McMillan C. *Mathematical Programming: An Introduction to the Design and Application of Optimal Decision Machines.* New York: John Wiley & Sons, Inc., 1970.

Simmonnard, M. *Linear Programming.* Englewood Cliffs, N.J.: Prentice-Hall, Inc., 1966.

EXERCISES

1. A taxi dispatcher has six idle taxis that can be sent to five customers who have just called for rides. Because the amount of the fare in each case is not known, the dispatcher decides to assign taxis to customers so that the total mileage of getting to these five customers is minimized. The dispatcher knows the current location of his taxis and the customers, and he determines the mileage between each taxi and customer to be:

		Customer				
		A	B	C	D	E
	1	30	10	2	15	18
	2	20	6	12	0	23
Taxi	3	18	20	5	9	29
	4	8	15	30	25	13
	5	19	18	35	12	7
	6	10	21	13	24	10

What assignment of taxis to customers should the dispatcher choose to minimize total mileage?

2. Four manufacturing plants must be built in four different geographic regions, with only one plant in each location. The costs of building each plant in the various locations are given below.

Cost in $ millions

		Location		
	1	2	3	4
1	60	51	32	32
2	48	X	37	43
3	39	26	X	33
4	40	X	51	30

Plant (rows: 1, 2, 3, 4)

In this matrix an "X" indicates that, because of the need for highly specialized labor, it is impractical to assign a plant to a particular location. Determine the least-cost allocation of plants to locations.

3. Three ships are to be unloaded at a dock in which four berths are available. By virtue of the ships' cargoes and the unloading facilities at the berths, different time periods are required for the unloading at each berth. These data are shown below:

		Ship	
	1	2	3
1	5	13	19
2	13	10	15
3	11	15	27
4	15	9	6

Berth (rows: 1, 2, 3, 4)

Thus, to unload ship 2 in berth 3 takes 15 days. Find the assignment of ships to berths which minimizes the total ship-days of unloading time.

4. The following four job lots have just arrived in your lathe department, which consists of four different types of lathes. How many standard hours* of work would you schedule on each machine to minimize costs?

	Cost per standard lathe hour — Lathe				Standard hours required
	A	B	C	D	
1	4	5	3	8	20
2	3	2	3	4	35
Jobs 3	2	1	2	3	45
4	1	6	2	5	20
Standard available hours	20	40	10	50	

5. A total of 400 railroad cars is available at five locations. A total of 350 cars is required at the four locations as indicated below. What redistribution plan would you use to minimize costs?

*Data must be converted to standard hours. This has been done for the matrix as shown above. Use the most effective lathe as a basis, say A. If lathe B is capable of producing at half the rate of A and lathe B has 80 actual hours available, then lathe B will have 40 standard hours available. If C can produce at 0.8 the rate of A and has 12.5 actual hours available then it will have 10 standard hours available, etc. The standard hours required for job 1 are the actual hours required if processed on lathe A. The costs shown in the matrix are for standard hours. For example, if lathe B produces at half the rate of lathe A and the cost per actual hour is 2.5, then the cost per standard hour is 5. These comments are included to guide the reader in formulating a problem of this nature. The transportation algorithm can be applied directly to the above matrix.

Transportation costs per car

		Destination 1	2	3	4	Availability
	1	4	2	6	7	50
	2	1	9	7	8	100
Source locations	3	2	8	5	2	60
	4	7	1	1	6	90
	5	4	9	4	1	100
Demand		75	75	50	150	

6. A company plans to produce three new chemical products at two existing plants to utilize some idle facility time. Each of the three new products requires approximately the same facility hours per pound. The company wants to produce to meet sales contracts. How much should they produce to minimize production cost?

Production cost (¢/lb.)

		Product 1	2	3	Capacity (1000lb./day)
Plant	A	6	5	6	20
	B	4	8	3	15
Contracted sales (1000lb./day)		5	10	8	

7. The Fly-By-Night Airline must decide on the amounts of jet fuel to purchase from three possible vendors. The airline refuels its aircraft regularly at the four airports it serves.

The oil companies have said that they can furnish up to the following amounts of fuel during the coming month: 250,000 gallons for Oil Company 1, 500,000 gallons for Oil Company 2 and 600,000 gallons for Oil Company 3. The required amount of jet fuel is 100,000 gallons at Airport 1, 200,000 gallons at Airport 2, 300,000 gallons at Airport 3 and 400,000 gallons at Airport 4.

When transportation costs are added to the bid price per gallon supplied, the combined cost per gallon for jet fuel from each vendor servicing a specific airport is shown in the table below.

	Company 1	Company 2	Company 3
Airport 1	12	9	10
Airport 2	10	11	14
Airport 3	8	11	13
Airport 4	11	13	9

(a) Formulate this situation as a linear programming problem, letting x_{ij} equal the amount of fuel sent by company j to airport i.

(b) By using Vogel's Approximation Method, solve this problem with the transportation algorithm.

8. Obtain an integer solution to the problem of Example 2, in Chapter 5, on page 124.

9. A certain company operates monthly excursions on the Tennessee River. Each month at least 7,500 passengers take the tour. Two types of pleasure boats are available in the company's fleet:

	Type 1	Type 2
Capacity, persons	2,000	1,000
Fuel, gallons per trip	12,000	7,000
Crew, employees	250	100

Total fuel consumption is highly important because the company has been allotted only 55,000 gallons per month. In addition, the total number of employees cannot exceed 900. The company expects to make $20,000 net profit per Type 1 pleasure boat used and $10,000 net profit per Type 2 boat.

Determine how many boats of each type are required to maximize net profit. Use cutting planes to find integer values for the two decision variables.

SEARCH PROCEDURES

INTRODUCTION

A pragmatic question often asked in mathematical programming courses is, "Where do you get these mathematical equations you use to optimize real-world problems?" Even though these mathematical models are applicable to many special problems, there are no neat, easily derived mathematical expressions for a majority of industrial problems.

To illustrate this notion, suppose an executive realizes that profit is a function of advertising expenditures where x_1 is television advertising dollars, x_2 is radio advertising dollars and x_3 is personal sales dollars. Although many executives can use their judgment and experience to speculate on the relationship between advertising expenditures and profits, the exact mathematical relationship between profits and these three variables is unknown. This same problem, not unique to business, is frequently encountered in scientific experimentation and engineering design work. For example, suppose quality control engineers in a food cannery discover that bacteria count in food is related to the percent of a particular additive, x. There are no scientific laws that enable them to express bacteria count as a function of x.

Many practitioners are unprepared to cope with this type of problem because their college background has prepared them for a logical world in which important relationships can be discovered and expressed in neat mathematical form. When the practitioner encounters this predicament he will search as best he knows how for the optimum.

The advertising executive will decide on dollar figures for his television and radio advertising and personal sales effort and will observe the effect on profit. Using his experience, he will then vary one or more of these expenditures and determine whether this increases or decreases profit. He will continue to evaluate the response function (i.e., profit) at different points, or values of the decision variables, until new points give no improvement in the response.

The biologist will run tests with different values of x and observe the bacteria counts. He will then use his initial results to select values of x for further tests. The testing will terminate when he finds a certain level of x at which small increases or decreases in the additive will give an increased bacteria count. In most cases, the searcher depends on a combination of intuitive judgment and luck to locate the optimum.

Practitioners quickly learn two principles that were not emphasized in their college training. First, if there is any way for a venture to fail it usually will. Second, there is a cost of evaluating the response at a particular point that may seriously restrict the search. The objective of this chapter is to teach the student to search for optimum values of decision variables in a prudent manner. Such an "organized" search minimizes the effect of "bad luck," thereby minimizing the number of times the response must be evaluated and the cost of needless experimentation.

One principle of searching is of vital importance. In any search the response must be evaluated at a certain number of points. The benefit of this effort is the location of an improved response. In actual practice there is a cost of evaluating the response. In some cases this cost may be trivial as in our example of bacteria counts. In other cases, a single point may easily cost thousands of dollars to evaluate, as in our advertising example. The investigator must always use judgment in investing in searching.

All of this does not mean that mathematical programming is not applicable to the real world. Mathematical expressions can be derived in many cases. Linear programming problems can be formulated from known costs, utilization factors, resources and so forth. Engineering and scientific laws and rules can be used to formulate precise mathematical objectives that may be linear or nonlinear. In some areas, experience with real-world operations may give sufficient background to develop an approximating function and its parameters. Response surface methodology may be used with experimental test results to develop an approximating function for the response surface.

In some cases, particularly in science and engineering, a mathematical expression can be derived from theory, but it is so complex that a solution may be impractical. The search procedures discussed in this chapter are suited to this situation. However, when the objective function is known, there are special applicable procedures that are more effective than those discussed here. The student should refer to a more advanced text for these procedures.

The search procedures presented in the first part of the chapter assume that the objective function is convex or concave. When the investigator locates a point at which minor changes in the decision variables give no improvement in response, he may conclude that this is the global optimum. The frustrating problem of multiple stationary points will be briefly discussed in the last section of this chapter.

FUNCTIONS OF

ONE VARIABLE

In many cases, an executive, an engineer or a scientist is interested in the effect that one decision variable will have on a response. As an example of this type of problem we may think of the decision variable, x, and the response, y, as any of the following:

Response, y	*Decision Variable, x*
Profit	$x for radio advertising
Tensile strength of a metal	x% of an alloy
Bacteria growth	x% of an additive
Inventory stockout cost	x units in buffer stock
Chemical yield	x units of a catalyst
Percent defectives produced	x intensity of lighting on an assembly line
Manufacturing system costs	x items as work in process at an assembly station in a manufacturing plant

The investigator will evaluate the response at a particular value of the decision variable and use these results to locate the next value of the decision variable for evaluation. The objective of this section is to illustrate this procedure graphically. As an example, we will consider the problem of determining what expenditure on radio advertising, x, will maximize profit, y. (The basic principles illustrated here are equally applicable to science and engineering.)

The response function is shown as a dashed line in Figure 7-1. This function is unknown to our advertising executive. It is assumed that this is a concave function as shown. If it is suspected that profit is not a concave (or convex) function of radio advertising, then the procedures discussed in the last section of this chapter should be applied. The region of search has been bounded to the x_L = $0 to x_U = $100,000 region since this has been predetermined as the

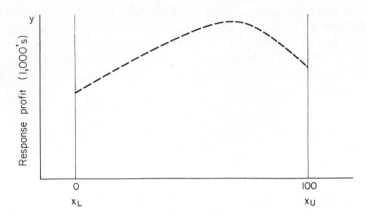

FIGURE 7-1   A Bounded Convex Function of One Variable

feasible range of expenditures, where x_L and x_U equal lower and upper bounds of the feasible interval, respectively.

Assume that we conduct a radio campaign for x_1 = $25,000 and realize a profit of $975,000 as shown in Figure 7-2. A second campaign of x_2 = $60,000 yields a profit of $1,200,000, also shown in Figure 7-2. We have assumed the objective function to be convex, thus the interval from x_L to x_1 can be eliminated. Since the function is concave, and $f(x_2)$ = $1,200,000 > f(x_1)$ = $975,000, it is impossible to have a value of $f(x)$ greater than $975,000 for $0 \leqslant x \leqslant$ $25,000. We now know that an expenditure between x_1 = $25,000 and x_U = $100,000 will give optimum profit. This is our inverval of uncertainty.

If a third campaign is conducted for x_3 = $90,000 and a profit of $1,300,000 results, as shown in Figure 7-3, then we may eliminate the interval x_1 = $25,000 to x_2 = $60,000 from further investigation.

This process is continued until the final interval of uncertainty is acceptable, or the investigator is unable to evaluate additional points because of constraints on such factors as time and money. If the investigator is very lucky, he may locate the optimum with only two or three points. However, it is not prudent management to depend on luck. Tactics for placing points should be carefully planned. One measure of a procedure's effectiveness is the maximum interval of uncertainty that could possibly remain after a certain number of response evaluations. It is desirable to make this interval as small as practical. Tactics that accomplish this are referred to as minimax procedures.

There are a variety of minimax procedures. If the observer does not know in advance exactly how many experiments he is going to conduct, as in the case of our advertising executive, the Golden Section is the most desirable procedure. Other procedures are more effective when the exact number of experiments is determined in advance, but the Golden Section is easy to use and provides

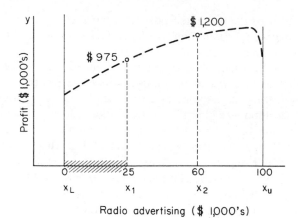

FIGURE 7-2 Two Initial Search Points

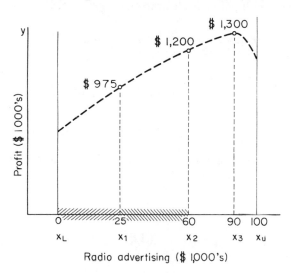

FIGURE 7-3 A Third Search Point

accurate results. The final interval of uncertainty with the Golden Section is at worst 17 percent longer than the best possible plan that may be developed when the number of experiments is determined in advance. The Golden Section is easily explained; its application is straightforward.

golden section

The Golden Section will be presented through the previous advertising example. Our advertising executive will use this plan for the systematic selection of points.

He is concerned with the maximum possible interval of uncertainty that could exist after evaluating a certain number of points. He knows that he may guess and locate the optimum with only two or three advertising campaigns if he is "lucky." However, he realizes that the interval of uncertainty can be very large if he is "unlucky," and he wishes to minimize the consequences of bad guesses. The cost of evaluating the response at a point is difficult to measure. The true cost is the actual difference in profits between those actually realized during the testing program and those that would have resulted during the same period if the executive had operated at a "guessed" value of x. This would include the cost of executive time required to evaluate the campaigns as well as lost sales.

The Golden Section is based on the constant 0.618034. For a discussion of this constant, see Wilde and Beightler (referenced at the end of this chapter). Search points are placed sequentially in the interval of uncertainty so they are symmetric and are at a point that is 0.618034 of the interval of uncertainty. The application of six search points is illustrated in Figure 7.4. Here we use 0.618 in our calculations.

The first two points are shown in Figure 7-4(a). The initial interval of uncertainty is $x_U - x_L$.

$$0.618 \, (x_U - x_L) = 0.618 \, (100.0 - 0.0) = 61.8$$

The first two points are placed symmetrically at a distance of 61.8 from each end of the interval of uncertainty.

$$x_1 = 0.0 + 61.8 = 61.8$$
$$x_2 = 100.0 - 61.8 = 38.2$$

Experiments are conducted at x_1 and x_2 and, since $f(x_1) = \$1,210,000 > f(x_2) = \$1,010,000$, from experimental results as shown in Figure 7-4(a), the interval from 0 to 38.2 can be eliminated. The interval of uncertainty becomes 38.2 to 100.0. This process of elimination is continued until the final interval of uncertainty is acceptable, or additional testing is not justifiable (i.e., our executive feels the point of diminishing returns has been reached).

The third point, x_3, is placed in this new interval at a distance of 0.618 $(100.0 - 38.2) = 38.2$ from the left end of the interval in Figure 7-4(b).

$$x_3 = 38.2 + 0.618 \, (100.0 - 38.2)$$
$$= 38.2 + \quad 38.2 = 76.4$$

At this stage of the search there are two points, 61.8 and 76.4 in the previous interval of uncertainty. Each of these points is a distance of 38.2 units from one end of the interval, and they are therefore symmetric. The resulting interval of uncertainty is $x_1 = 61.8$ to $x_U = 100.0$ since $f(x_3) = \$1,300,000 > f(x_1) = \$1,210,000$, as shown in Figure 7-4(b). The procedure is continued in Figure 7-4(c), with a fourth advertising campaign.

$$x_4 = 61.8 + 0.618\,(100.0 - 61.8)$$

$$= 61.8 + 23.6 = 85.4$$

$$f(x_4) = \$1,400,000 > f(x_3) = \$1,300,000$$

The resulting interval of uncertainty is $x_3 = 76.4$ to $x_U = 100.0$. The fifth advertising campaign is conducted at x_5, where

$$x_5 = 76.4 + 0.618\,(100.0 - 76.4)$$

$$= 76.4 + 14.6 = 91.0$$

$$f(x_5) = \$1,280,000 < f(x_4) = \$1,400,000$$

The resulting interval of uncertainty, as shown in Figure 7-4(d), is $x_3 = 76.4$ to $x_5 = 91.0$.

The final point, x_6, is placed in Figure 7-4(e) and is located at

$$x_6 = 91.0 - 0.618\,(91.0 - 76.4)$$

$$= 91.0 - 9.0 = 82.0$$

$$f(x_6) = \$1,370,000 < f(x_4) = \$1,400,000$$

and the final interval of uncertainty is from x_6 to $x_5 = 91.0 - 82.0 = 9.0$.

Our advertising executive chooses to terminate the search at this point. The final interval after six searches is $(91.0 - 82.0)/100.0 = 9$ percent of the initial interval. The executive will spend $85,000 per year on radio advertising, which will result in a profit of $1,400,000 per year assuming time does not alter the response pattern. (The reader may construct a concave function with the same initial interval of uncertainty and evaluate six points. Your final interval should be 9.0 since the final interval is independent of the function.)

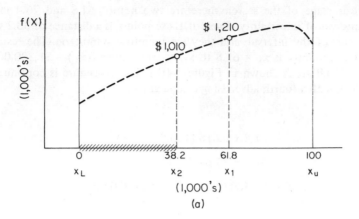

(a)

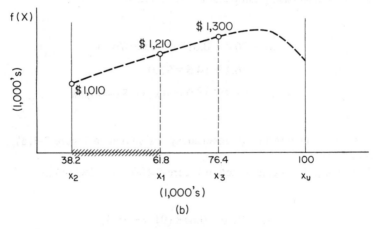

(b)

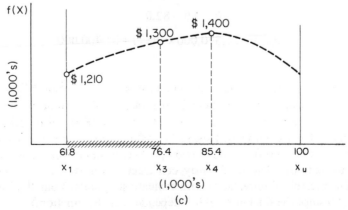

(c)

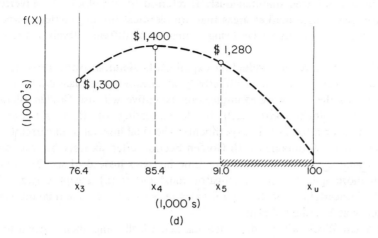

(d)

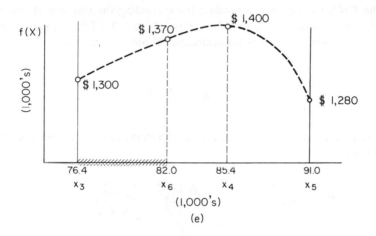

(e)

FIGURE 7-4 Application of the Golden Section

golden block

In the Golden Section, one experiment is conducted, and the results are analyzed and used to locate the next experiment. This one-experiment-at-a-time approach makes the most effective use of the information from each experiment. In some cases, however, time is the crucial factor. One experiment, such as our advertising example, may take months to complete. To shorten the search time some experiments may be taken simultaneously. A group of experiments conducted simultaneously is referred to as a block. Our advertising executive may locate market areas that are identical for all practical purposes, and conduct several radio advertising campaigns at different levels at the same time.

This procedure of conducting experiments simultaneously does not use information from each test as effectively as does a sequential (one-at-a-time) procedure. In the following example, our executive will use Golden Block to conduct two blocks, with each block consisting of three simultaneous experiments, for a total of six experiments. The final interval of uncertainty will be greater than 9.0 obtained with Golden Section after six tests, but the results will be available after two experiment time intervals instead of six. To obtain a solution more quickly, the investigator must either (a) accept a larger final interval of uncertainty, or (b) pay for extra tests to get the same final interval of uncertainty as a sequential plan.

Golden Block will be described mathematically and then applied to the advertising example.

The Golden Block is a procedure for evaluating the response at multiple points simultaneously. An odd number of points, b = 3,5,7,9..., is used in each block of this procedure. The first measurement should be taken at

$$x_1{}^1 = \frac{x_U - x_L}{W_b} + x_L$$

A subscript denotes the point number and a superscript indicates the block number where

$$W_b = 1/2 \left[\frac{(b+1)}{2} + \sqrt{\left(\frac{b+1}{2}\right)^2 + 4\left(\frac{b+1}{2}\right)} \right]$$

and x_U and x_L are the upper and lower bounds of the search region.

The second point should be at

$$x_2{}^1 = \frac{(x_U - x_L)}{\left(\frac{b+1}{2}\right)} + x_L$$

and the rest at

$$x_q{}^1 = x_{q-2}{}^1 + \frac{(x_U - x_L)}{\left(\frac{b+1}{2}\right)} \quad \text{for } q = 3, \dots, b$$

The interval of uncertainty remaining after the first block is I_1 (I_j is the interval of uncertainty after the jth block).

$$I_1 = \frac{x_U - x_L}{\left(\frac{b+1}{2}\right)}$$

After the second block of measurements the interval of uncertainty will be

$$I_2 = \frac{I_1}{W_b}$$

In fact, the interval of uncertainty will be reduced by a factor of $1/W_b$ for each block after the first block.

$$I_j = \frac{I_{j-1}}{W_b}$$

After the first block, this interval of uncertainty may be used to locate points. Every point in the interval of uncertainty should be located so that the point to the left is I_{j-1}/W_b units from the point to the right. The Golden Block method is illustrated in terms of the example used in the previous section.

Our advertising executive has located three market areas and wishes to test the effect of radio advertising on profit. The sales and advertising costs from each market area are used to calculate the total profit that would result if that level of advertising were used in all areas. The first three points are chosen by direct application of the formula and are illustrated in Figure 7-5(a).

$$b = 3$$

$$W_b = 1/2\left[\left(\frac{3+1}{2}\right) + \sqrt{\left(\frac{3+1}{2}\right)^2 + 4\left(\frac{3+1}{2}\right)}\,\right]$$

$$= 2.73$$

$$x_1{}^1 = \frac{x_U - x_L}{W_b} + x_L = \frac{100.0 - 0.0}{2.73} + 0.0 = 36.6$$

$$x_2{}^1 = \frac{x_U - x_2}{\left(\frac{b+1}{2}\right)} + x_L = \frac{100.0 - 0.0}{\left(\frac{3+1}{2}\right)} + 0.0 = 50.0$$

$$x_3{}^1 = x_1 + \frac{x_u - x_L}{\left(\frac{b+1}{2}\right)} = 36.6 + \frac{100.0 - 0.0}{\left(\frac{3+1}{2}\right)} = 86.6$$

It can be seen from Figure 7-5(a) that the interval of uncertainty after the first block is $I_1 = 100.0 - 50.0 = 50.0$, which may be calculated by

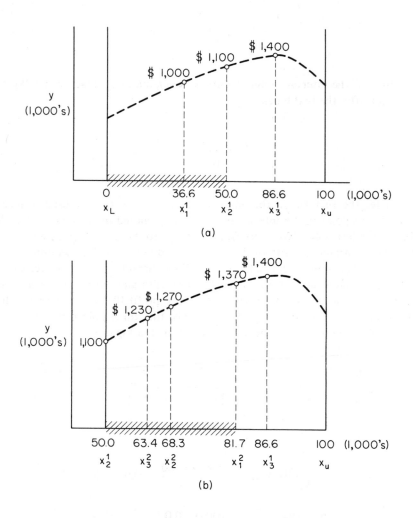

FIGURE 7-5   Application of the Golden Block

$$I_1 = \frac{x_U - x_L}{\left(\frac{b+1}{2}\right)} = \frac{100.0 - 0.0}{\left(\frac{3+1}{2}\right)} = 50.0$$

After the second block is applied, the interval of uncertainty will be

$$I_2 = \frac{I_1}{W_b} = \frac{50.0}{2.73} = 18.3$$

The point in I_1 after the first block is at $x_3^1 = 86.6$. The first point in the second block, x_1^-, is placed at $x_U - I_2 = 100.0 - 18.3 = 81.7$. The second point, x_2^2, at $x_3^1 - I_2 = 86.6 - 18.3 = 68.3$, and the third point x_3^2, at $x_1^2 - I_2 = 81.7 - 18.3 = 63.4$. It can be seen in Figure 7-5(b) that for every point from x_2^2 to x_3^1, the point on its immediate left is 18.3 units from the point on its immediate right. The final interval of uncertainty *must* be 18.3. The actual final interval is $x_1^2 = 81.7$ to $x_U = 100$.

With six points using Golden Section, the final interval of uncertainty is 9.0, which is about half that obtained with six points using Golden Block. If time is available, it is much more efficient to make experiments sequentially. If each point requires one month to evaluate, then an interval of 18.3 is reached after two months with Golden Block using b = 3 with two blocks, whereas Golden Section has an interval of uncertainty of 38.2 after two months. If time is of the essence, Golden Block is clearly more attractive.

When b increases, more experiments will be taken in each block, and the effectiveness of the information from each experiment will diminish. The experimenter should evaluate different search plans before beginning experimentation. The number of experiments and the final interval of uncertainty associated with each plan should be calculated.

FUNCTIONS OF MULTIPLE VARIABLES

In many realistic problems, the search is complicated when the response is clearly a function of multiple variables. An advertising executive may use several different media for advertising. The effect of expenditures for television advertising is dependent on expenditures for the personal sales effort. These two decision variables cannot be considered separately in this case. As an example of this type of problem, we may think of a response, Y, and the vector of decision variables, X, as any of the following:

Response, Y	*Decision Variables, X*
Profit	$\$x_1$ for advertising $\$x_2$ for personal sales
Probability of surviving surgery	$x_1\%$ of gas 1, $x_2\%$ of gas 2, and $x_3\%$ of gas 3 in anesthesia
Inventory cost	x_1 units in buffer stock and x_2 units in reorder quantity
Plant growth	x_1 lb. per acre of fertilizer 1 and x_2 lb. per acre of fertilizer 2
Manufacturing system costs	x_1 items in work-in-process and x_2 material handlers
Chemical yield	x_1 °C temperature, and x_2 psi

The search procedures used with multivariable problems are entirely different from those applied to single variable problems. With a single variable, the search procedures involve the elimination of line segments. Search procedures based on elimination of areas and volumes have proved extremely disappointing with multivariable problems. Functions used as examples in this section will involve two variables since these can be illustrated with contour mappings as shown in Figure 7-6. The student should refer to Chapter 2 for an introduction to contour mappings. The search procedures discussed here can be easily extended to functions involving three or more variables.

Our advertising executive may be viewed as a mountain climber who is attempting to locate the top of a hill. In our analogy, x_1, advertising expenditures and x_2, personal sales dollars, are the coordinates of the map. The elevation at a particular value of x_1 and x_2 is the profit associated with that decision point. The weather is foggy, so the climber cannot see the outline of the hill. All the climber can do is to proceed to some coordinate on the map and measure the elevation at that location. He will use previous results to guide himself when stumbling through the fog.

When he suspects that he has located the hilltop, he will move out short distances and measure the elevation. If these elevations are less desirable, he will assume that he has located the point of highest elevation. In the following examples, detailed calculations are not utilized. The student should study the graphical illustrations and imagine himself to be the mountain climber (who in his spare time is an advertising executive). Each small circle represents an

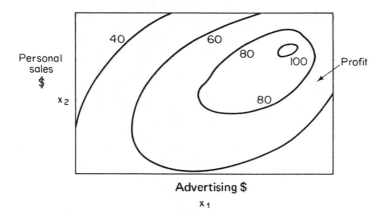

FIGURE 7-6   A Contour Mapping of a Bounded Function of Two Variables

evaluation of the response using the associated decision variables. These are numbered to indicate the sequence in which they are evaluated.

sectioning

Sectioning is a multivariable search procedure that is straightforward and relatively simple to apply. Many investigators not familiar with formal search procedures will intuitively use some variation of this procedure.

 With sectioning all variables are held constant but one. Incremental steps are taken in this one variable to locate an improved response. If an incremental step in one direction yields no improvement then an incremental step is taken in the other direction. When these increments give no improvement in response, the variable being tested is held constant and another is varied in the same manner. This procedure will terminate when these increments give no improvement in any variable.

 Sectioning is illustrated in Figure 7-7. The points are numbered in the order in which they are evaluated. With a function such as shown in Figure 7-7, this procedure is effective. When narrow ridges are present sectioning is less effective. This procedure tends to terminate prematurely at narrow ridges, as shown in Figure 7-8.

 Sectioning is often initially applied with large incremental steps. When this procedure terminates, a second sectioning is applied with much smaller incremental steps. This plan is devised to minimize the total number of response evaluations. An application is illustrated in Figure 7-9.

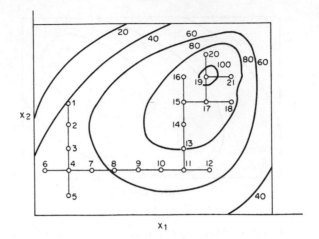

FIGURE 7-7   Sectioning

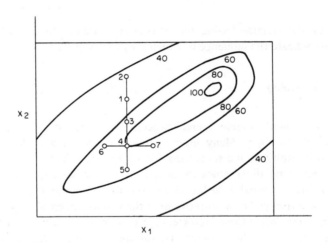

FIGURE 7-8   Sectioning Terminated by a Ridge

pattern search

A pattern search is illustrated in Figure 7-10. This illustration will be referred to frequently in the subsequent discussion. In pattern search, a base point, B_1, is evaluated initially. An exploratory point is then evaluated at some small incremental distance in one variable. This is referred to as a *perturbation*. If this is not more desirable than the base point then a perturbation is taken in the other direction in the same variable. This is illustrated by points 1 and 2 of Figure 7-10. (The entire pattern search is illustrated so the reader

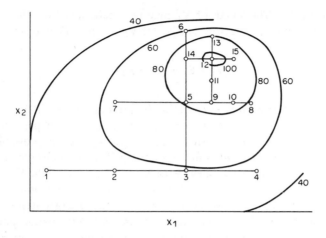

FIGURE 7-9   A Sectioning Plan Using Different Incremental Step Sizes

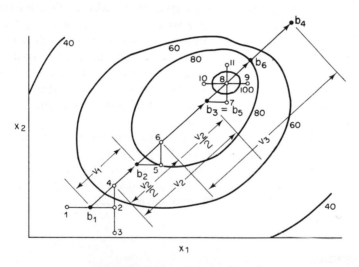

FIGURE 7-10   Pattern Search

can visualize an entire pattern search applied to a problem. Although this appears complex initially, the student will realize that pattern search is exceptionally simple after relating the following discussion to Figure 7-10.) If perturbations in both directions yield no improvement, then similar perturbations are made in another variable. If an improvement is found, the investigator will make perturbations about the new point (points 3 and 4 of Figure 7-10).

A vector of length V_1 is then drawn from the original base point, B_1, through the last desirable perturbation (point 4 of Figure 7-10) to locate the next base point, B_2. The length of V_1 may be determined from the experience and judgment of the investigator. The vector V_1 is often taken as the following:

$$V_1 = P_1 + (P_1 - B_1)$$

where P_1 is the last successful perturbation (point 4 in Figure 7-10).

The perturbation process is repeated about B_2. Initial perturbations in each variable may be made in the direction of expected improvement. This results in the location of point 6 in Figure 7-10. If these results give a direction of improvement that is not colinear with V_1, a vector of length V_1 should be used to locate the next base point. If these results give a direction of improvement colinear with V_1, as shown in Figure 7-10, then a larger step size may be taken. This larger step, V_2, is determined by the last perturbation point about the present base point (point 6) and the last perturbation point about the previous base point (point 4). The step V_2 is twice the distance between these two points. Repeating this procedure gives continually accelerating step sizes that have the capability of converging rapidly to a remote optimum. Eventually this will result in a step size that "overshoots" the optimum, as shown by V_3 of Figure 7-10. When a base point such as B_4 gives no improvement over the previous base point, the search should be resumed at the previous base point using a step size of V_1. In Figure 7-10, this results in B_5. Perturbations about B_5 are successful in locating the optimum. Perturbations about point 8 yield no improvement and the procedure terminates.

If the investigator suspects that the pattern search has terminated at a narrow ridge, he may rotate the axis of the pattern by 45 degrees and continue. The pattern search illustrated in Figure 7-11 will terminate at B_2 unless the axis is rotated. Points 7 and 8 are the initial perturbations on the rotated axis. By rotating the axis of perturbation the optimum is located.

MULTIMODAL

FUNCTIONS

A multidmodal function of one variable is illustrated in Figure 7-13, and of two variables, in Figure 7-12. The student should study these graphs from the viewpoint of an executive, scientist or engineer who is trying to locate the optimum, by searching. When you examine these functions in detail from this perspective, you will realize what a frustrating problem this often presents to the investigator. A searcher should always consider the possibility that his problem is multimodal. When a problem can be identified as being multimodal, much of the battle has been won, since the procedures recommended here may be applied.

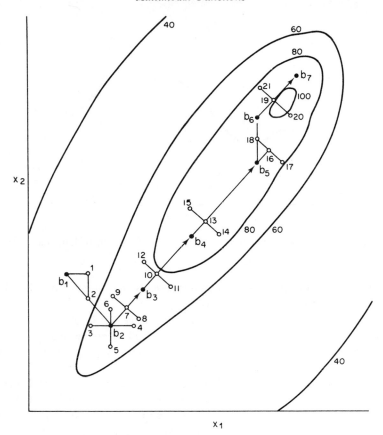

FIGURE 7-11   Pattern Search with Rotated Axis

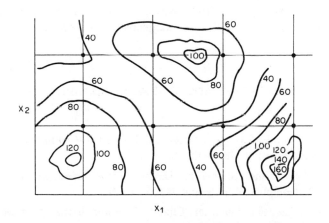

FIGURE 7-12   A Multimodal Continuous Function of Two Variables

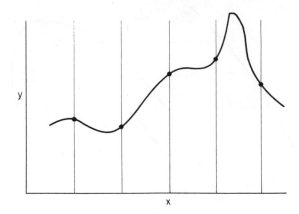

FIGURE 7-13   A Multimodal Continuous Function of One Variable

The multimodal problem most frequently defeats the investigator when he refuses to admit that his problem is multimodal.

A procedure does not exist that will guarantee the location of the global optimum of a multimodal continuous function, as illustrated in Figures 7-12 and 7-13. One approach to optimization is to superimpose a grid over the region of search as shown in Figures 7-12 and 7-13.

In the multidimensional problem, as shown in Figure 7-12, the response is evaluated at each vertex of the grid. The fineness of the grid will depend on the cost of evaluating the response relative to potential savings through improvement in response. Search procedures may be initiated from the vertices with the best response.

In the unidimensional problem, as shown in Figure 7-13, the response is evaluated at each grid point. The investigator will use his judgment and experience in examining results to subjectively narrow the interval of uncertainty.

SUGGESTED ADDITIONAL READINGS

Beveridge, Gordon S.G., and Schechter, Robert S. *Optimization: Theory and Practice*. New York: McGraw-Hill, 1970.

Cooper, Leon, and Steinberg, David. *Introduction to Methods of Optimization*. Philadelphia: W.B. Saunders Co., 1970.

Wilde, Douglass, and Beightler, Charles S. *Foundations of Optimization*. Englewood Cliffs, N.J.: Prentice-Hall, Inc., 1967.

EXERCISES

In the following exercises a real situation is described and a mathematical function is given for the response. To evaluate the response for a particular value of a decision variable, merely evaluate the mathematical function. In many cases when search procedures are applied to real problems, a mathematical expression of the response is not available. The response is evaluated through expensive testing programs.

1. Apply Golden Section and Golden Block to each of the following and compare their performances. Unless otherwise specified, apply Golden Section five times, and use two blocks of $b = 3$ and/or two blocks of $b = 5$ with Golden Block.

   (a) Let x be the average number of units of work held as a planned buffer stock at a work station. The total cost of worker idle time, inventory costs and defective production is given by $f(x) = x^2 - 10x + 25$ where $f(x)$ is in \$100/day. Assume x is a continuous variable for purposes of searching. Search over the interval 0 to 20.

   (b) Solve the problem in (a), but search over the interval of 0 to 100 instead.

   (c) Let x be thousands of dollars spent on advertising. Sales, in thousands of dollars per year, is given by $f(x) = 50x - 10$. Search over the interval of 0 to 50, to maximize sales.

   (d) Let x be the percent of alloy used in a metal. The tensile strength of a rod made from this alloy is given by $f(x) = -x^2 + 20x - 14$ where $f(x)$ is in thousands of pounds. Maximize its strength by searching from x $= 1.0$ to 11.0 percent.

2. In the following problems, use your judgment as to step sizes with sectioning and perturbation sizes with pattern search. You may revise these after beginning the search if experience indicates another value would be preferable. Apply sectioning and pattern search to each of the following:

   (a) Let x_1 be thousands of dollars spent for television advertising and x_2, thousands of dollars for magazine advertising. Corporate profit in thousands of dollars is given by

   $$f(x) = 200 - (50 - x_1)^2 - (2x_2 - x_1)^2$$

(i)   Beginning at x = [0,0]

(ii) Beginning at x = [20,20]

Search for values of x that maximize corporate profit.

(b) Let x_1 be the percent of additive 1 and x_2, the percentage of additive 2 at a cannery. Bacteria count in the product is given by

$$f(x) = 2 - 10x_1{}^2 + 20x_2{}^2 - 20x_1 x_2$$

(i)   Beginning at [10,10]

(ii) Beginning at [1,5]

Find the amount of each additive that minimizes bacteria count. Neither x_1 nor x_2 can be less than 1.0.

APPENDIX

CASE STUDIES

The case studies in this Appendix illustrate the material presented in the respective chapters as indicated below:

preface

"How They're Planning OR (Operations Research) at the Top," by Efraim Turban.

chapter 5—the
Simplex algorithm

"Linear Programming Guides Parts Buyers," by John L. Simpson and Dante J. Pellei.

"Optimum Press Loadings in the Book Manufacturing Industry," by Charles H. Aikens III.

"Introduction to Linear Programming for Production," by J. Frank Sharp.

chapter 6—special
forms of linear
programming

"Linear Programming Without the Math," by Edward Cochran.

"Linear Programming to Aid Resource Allocation in R&D," by A.G. Lockett and A.E. Gear.

HOW THEY'RE

PLANNING OR

AT THE TOP*

Abstract: This survey of 107 of the largest corporations in the United States shows how Operations Research is becoming an integral part of their corporate level activities.

Top management in many corporations is puzzling over the important yet difficult problem of how to introduce into their regular operations all the modern management tools and techniques of management science, operations research, and computer technology. While these techniques are spreading at plant and divisional levels for various operational decisions, there is hesitation and caution in the use of such techniques at the top level. Many executives think management is an art rather than a science, and they view management science activities as a waste at the top level. More progressive managers have tried these techniques and are using them with various degrees of success.

The existing departments appear under various names. Thirty-eight percent are called operations research departments; 18 percent, management science; 16 percent, systems; and 11 percent, corporate planning. Some other interesting names are: model development, scientific system development, operations analysis or evaluation, and business research.

But whatever it is called, top managers are confronted with a new problem—should it be used in their corporate level activities and if so, in what form and to what extent? Most of the largest US corporations have already made the decision to apply operations research to their corporate level operations. The crucial question that remains is—how?

First, the manager will want to know the advisability of establishing a special OR department at the corporate headquarters. The survey shows that 44 percent of the large corporations have such a department, and on the average, companies that have a special OR department are about 50 percent larger than those who do not. The average rate of sales growth in both groups is the same but the growth in terms of earnings per share is higher (12.4 vs. 8.7 percent) in the group that does not have an OR department. A probable explanation is that the size of the company affects the earnings per share rather than the existence of an OR department. And, just as sick people go to a doctor, so the companies with the lower rate of growth in earnings per share are seeking the aid of operations research.

*By Efraim Turban. Reprinted by permission of the publisher from *The Journal of Industrial Engineering*, December 1969, pp. 16-20.

Another interesting difference between the groups is their technological development, measured as the percent of the present product mix that is more than five years old. In the case of the group with the OR department, only 65 percent of its present product mix was being produced five years ago. The group that does not have such a department was producing 82 percent of its present product mix five years ago. (Ninety one percent, if corporations planning to establish a department are excluded.) The analysis shows also that OR departments are less common among conglomerates than among industrial corporations which operate mainly in one industry. An explanation may be that most conglomerates are relatively young and still in a formative process. On the other hand, a large portion of the specialized corporations either have such a department or plan to introduce one soon.

Figure 1 illustrates the growth of OR departments in the past as well as the projected growth. It is reasonable to predict that by the mid-70's about two-thirds of the large US corporations will have special OR departments at their corporate headquarters. With such an impressive growth, it is pertinent to ask why some companies do not have, or do not plan to have, such a department. Only two companies surveyed have actually abolished an OR department because they considered it an unsuccessful experiment. The majority of the other companies that do not have a special department report either that they are conducting operations research at various headquarters departments, or delegating such activities to the divisional level. Some companies have mentioned the formation of special task forces or committees to attack specific problems, while others have admitted that top management is not ready for such a progressive step. Those companies that distribute OR activities among various departments of headquarters, or those that delegate them to the divisional level, try to justify this as a cost saving and claim that this arrangement makes better use of manpower. The question remains as to how effective such second-priority activities are. If the function is delegated to a division, then corporate projects involving more than one division are probably not carried out effectively. If the function is given to another headquarter department there is a danger of low efficiency as explained by Gresham's law of planning.

The importance of the OR department can be evaluated through the analysis of its status. It seems that top management gives high recognition and authority to this department. Most of the departments report directly to a vice president, and in some cases, the director of the department is a vice president himself. The precise breakdown shows that 23 percent of the departments report to the vice president or director of systems, 21 percent report to the vice president of services, 17 percent to the vice president of planning and development, 12 percent to the president, 12 percent to the comptroller, 6 percent to the vice president of accounting, and 6 percent to the vice president of finance.

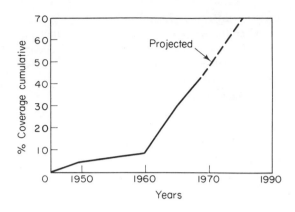

FIGURE 1   How the Percentage of Companies with OR Departments Is Growing

who mans the OR department?

Apart from the largest 10 percent, OR departments are very small, with an average of five professionals and 1¼ clerical and secretarial staff. Even if the few large departments are included, the average only goes up to about 10 professionals. One of the explanations for the small size is that the department's major objective is to increase efficiency and, as such, it should be an example of high quality with small quantity. This tendency to keep the department small probably means that the future rate of growth will be slower than that of the past.

The OR department is characterized by young and highly educated professional people. The average age of the department manager is forty, while the average age for the professional staff is thirty-three. Figure 2 gives the age distribution of a typical department. Figure 3 shows the educational level, which in industrial corporations was found to be higher than that in service corporations.

Operations research involves a multidisciplinary approach, so it is not surprising to find a diversified background among the professional staff. These range from engineering to historical studies. Also typical is the distribution of the major field of study among the bachelor's, master's, and doctorate levels. The most common fields of study on the undergraduate level are mathematics and statistics (about 26 percent), engineering (about 28 percent) while business education amounts to 14 percent. At the master's level, a strong shift is seen. About 26 percent of those with master's degrees have an MBA and about 24 percent have an MS in operations research or management science. The number

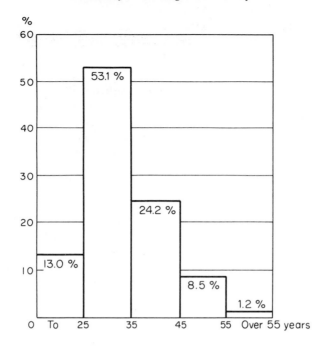

FIGURE 2 The Bulk of Members of an OR Department Is Under 35 Years Old.

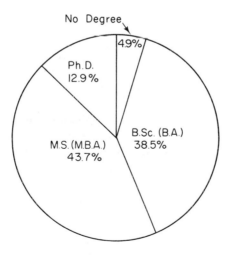

FIGURE 3 How Educational Levels Are Distributed Among Members of an OR Department

of mathematicians and statisticians decreases to about 16 percent and the engineers almost completely disappear (except for industrial engineers at about 7 percent). This level also includes some economists (about 6 percent).

On the doctoral level, 32 percent have a specific operations research or management science degree, 21 percent are mathematicians or statisticians, 23 percent are chemists and chemical engineers, and 14 percent are economists.

The directors of the OR department are on an even higher educational level—32 percent have doctoral degrees and 44 percent have master's degrees. The professional background of the director is mainly in operations research or management science (35 percent), and in engineering (29 percent).

The OR department is a young department, so it is not surprising to find that the average tenure of the professional employee is only four short years. This average stay can also be explained by the fact that several companies are using this department as a training school for top-line and staff positions. About 35 percent of the professionals who leave the department are promoted to a staff position and 19 percent are promoted to a line position in the same company. To replace the manpower losses caused by the high rate of turnover, recruiting is done directly at the college level in about 46 percent of the cases. In the other 54 percent, an experienced staff is preferred.

Finally, since the department is usually small, it is often necessary to use outside consultants. About 60 percent of the companies have made use of consulting firms and about 86 percent of these believe that the consulting costs were justified.

what does the
department do?

One question which the top executive usually puts to the management scientist is, "What will the OR department do at the corporate headquarter?" The most general answer to this question is that the management scientist should help the executive to make better decisions in all areas, but the practical executive will want examples. The survey recorded about 500 projects, from simple inventory problems to complex projects involving the simulation of the entire industry or social systems. Figure 4 lists some of the most frequently mentioned projects. The survey attempted to find trends in the OR activities by asking companies to list their past, present, and future projects. Generally, there is a clear trend to more sophisticated projects as shown in Figure 4.

One exigency which justifies OR activities at the corporate level is the need to carry out projects involving several divisions, or several plants belonging to different divisions. Such activities cannot be carried out by plant or even divisional OR units. However, corporate OR activities are not limited to the corporate level and the OR unit can carry out projects involving several plants or divisions. In several cases, during the course of analyzing a corporate level

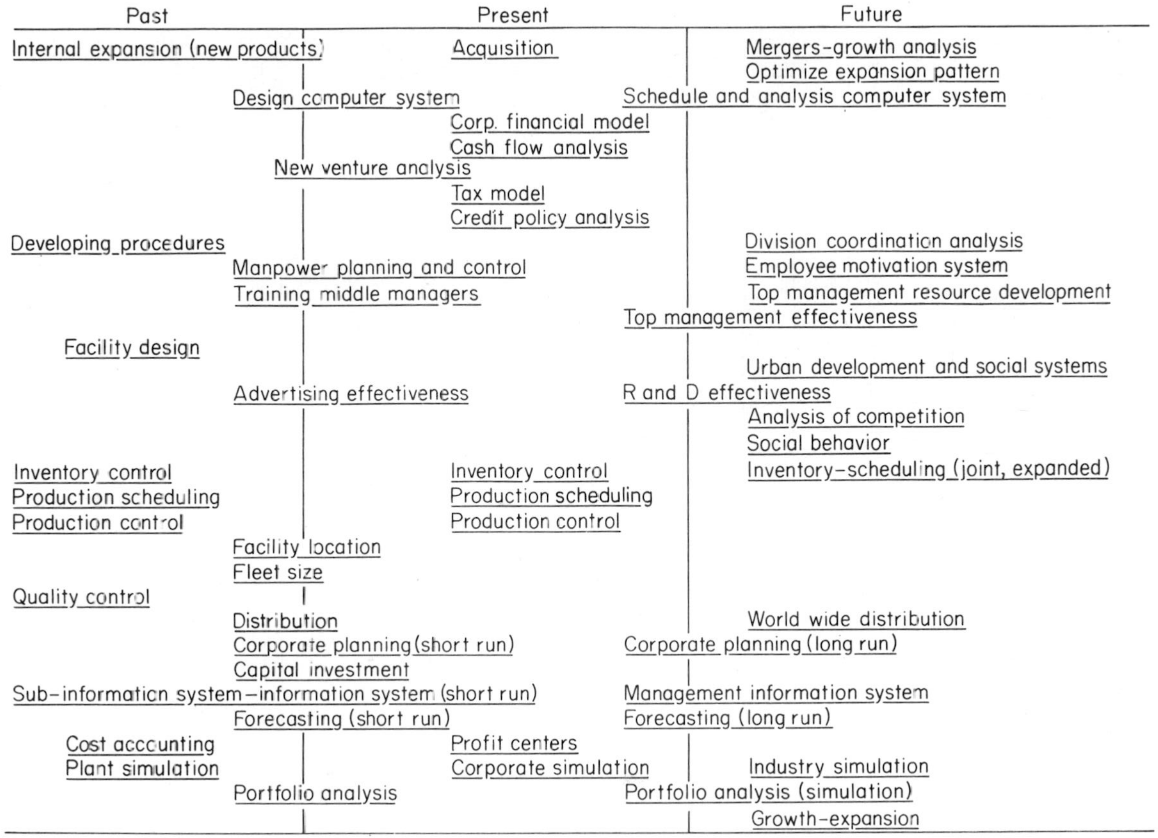

Past   Present   Future

Internal expansion (new products)
Acquisition
Mergers–growth analysis
Optimize expansion pattern
Design computer system
Schedule and analysis computer system
Corp. financial model
Cash flow analysis
New venture analysis
Tax model
Credit policy analysis
Division coordination analysis
Employee motivation system
Top management resource development
Developing procedures
Manpower planning and control
Training middle managers
Top management effectiveness
Facility design
Urban development and social systems
Advertising effectiveness
R and D effectiveness
Analysis of competition
Social behavior
Inventory control
Production scheduling
Production control
Inventory–scheduling (joint, expanded)
Inventory control
Production scheduling
Production control
Quality control
Facility location
Fleet size
World wide distribution
Distribution
Corporate planning (short run)
Capital investment
Corporate planning (long run)
Sub-information system–information system (short run)
Management information system
Forecasting (long run)
Profit centers
Corporate simulation
Forecasting (short run)
Cost accounting
Plant simulation
Industry simulation
Portfolio analysis (simulation)
Portfolio analysis
Growth–expansion

FIGURE 4   Money-Saving Activities That OR Departments Do—Past, Present, and Future

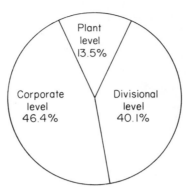

FIGURE 5 How OR Projects Are Distributed Between the Various Levels in the Companies

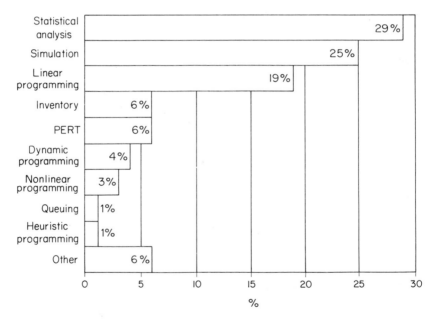

FIGURE 6 Most Operations Research Problems Are Solved by Statistical Analysis, Simulation, or Linear Programming. Only a Small Proportion Are Solved by Other Established Techniques or by Techniques Developed Specifically for the Problem

problem, the team must solve a plant problem. In some cases, this can be done by the local OR force; in others, it is done by the corporate OR department. Figure 5 shows the distribution of projects at the various corporate levels.

The average project lasts about 10 months and involves 2½ researchers. The OR personnel use the computer quite frequently. Most often they use the simplest OR techniques (such as simulation, linear programming, and statistical analysis). As one management scientist said: "We use simple models for the simple-minded managers, since 50 percent of our job is to sell what we do in the other 50 percent." Figure 6 shows the frequency of use of these techniques.

Last, but not least, is the problem of implementing the results of a project. Many OR scientists, especially at plant and divisional level, used to complain that a large portion of their projects was never implemented. This survey shows that today about two-thirds of all projects are mostly or completely implemented. This fact reflects the high status of the department and the strong support given to it by top management.

REFERENCES

1. Murray, G. L., "Scientific Vs. Practical Management: A Pragmatic Approach," *Management Services*, Jan-Feb 1967.

2. Drucker, P., "Management Science and the Manager," *Management Science*, Volume 1, 1954.

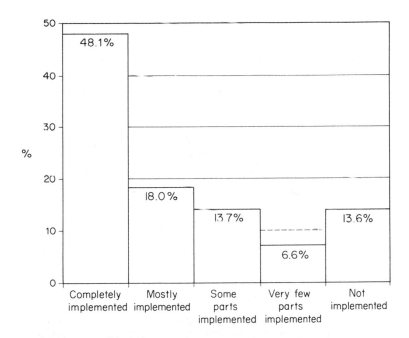

FIGURE 7   Today, Only 13.6 Percent of OR Projects Are Implemented

LINEAR

PROGRAMMING

GUIDES PARTS

BUYERS*

Abstract: A manufacturer's buyers of product components must contend with several variables—multiple suppliers (each with different prices and capacities) and monthly changes in product mix. Linear programming is used to determine least cost schedules for the manufacturer in this case study.

Successful solution of intricate purchasing problems has enabled the Diesel Engine Products Department of General Electric Company to obtain the lowest cost of parts regardless of product mix. The continued use of linear programming enabled the Department to minimize monthly costs of purchased parts from multiple suppliers, each with limited capacity, while the demand for individual parts has fluctuated.

The Diesel Engine Products Department produces diesel engines in the 1500 to 4000 horsepower range. In order to maintain a proper balance of inventories, parts and stock for components are ordered on a periodic basis. This problem is not concerned with how much stock to order, but who to order it from when each supplier can furnish only a portion of the total requirements.

For example, since no one foundry is geared to supply all the monthly requirements for a specific part, the Department has several suppliers, each with limited capacity. The size of order each foundry gets depends on the quantities of each part required each month, and the relationship of its costs to the total costs of a group of parts furnished by that particular foundry and several others. Obviously, the prices of the foundries vary for each of the parts they are able to supply.

To simplify, we will limit the illustration to three suppliers, A, B and C, and will list the cost per part from each supplier as shown in the following table, assuming the part will be different for each engine size.

	Cost per part for engine size		
Supplier	*8 Cylinder*	*12 Cylinder*	*16 Cylinder*
A	$647	$785	$842
B	$652	$762	$878
C	$683	$728	$893

*By John L. Simpson and Dante J. Pellei. Reprinted by permission of the publisher from *The Journal of Industrial Engineering*, April 1973, pp. 38-40.

We will let the variable X_n equal the monthly quantity of a part for a specified engine size from a specified supplier. These variables are assigned as follows:

Supplier	Monthly quantities of parts by engine size		
	8 Cylinder	*12 Cylinder*	*16 Cylinder*
A	X_1	X_2	X_3
B	X_4	X_5	X_6
C	X_7	X_8	X_9

The purpose is to order the best mix of parts from each supplier every month to minimize the monthly costs to the Department.

Supplier	Monthly costs of parts by engine size		
	8 Cylinder	*12 Cylinder*	*16 Cylinder*
A	$\$647(X_1)$	$\$785(X_2)$	$\$842(X_3)$
B	$\$652(X_4)$	$\$762(X_5)$	$\$878(X_6)$
C	$\$683(X_7)$	$\$728(X_8)$	$\$893(X_9)$

The total of the costs in the preceding table is to be minimized each month.

Now let us develop some constraints, in terms of the variables listed above, based on facts.

Supplier limitations	*Constraint*
A has a capacity limitation of 24 units per month.	$X_1 + X_2 + X_3 \leqslant 24$
B has a capacity limitation of 4 16-cylinder parts per month.	$X_6 \leqslant 4$
B has a capacity limitation of 30 units per month.	$X_4 + X_5 + X_6 \leqslant 30$
B has a contract which insures he will receive an order for at least 26 units per month.	$X_4 + X_5 + X_6 \geqslant 26$

(continued)

Supplier limitations	*Constraint*
C is having problems with tooling this month on his 12-cylinder units and cannot produce this size.	$X_8 = 0$
C measures this capacity in equivalent 8-cylinder units, can make 25 equivalent 8-cylinder units per month. He ratios the equivalent as follows: 1 16-cylinder unit = 1.8 equivalent 8-cylinder unit; 1 12-cylinder unit = 1.4 equivalent 8-cylinder units.	$1.0X_7 + 1.4X_8 + 1.8X_9 \leqslant 25$

To demonstrate the calculation of materials requirements for a month, let us assume that the total number of units required are:

$$\text{8-cylinder sizes} = X_1 + X_4 + X_7 = 21$$

$$\text{12-cylinder sizes} = X_2 + X_5 + X_8 = 27$$

$$\text{16-cylinder sizes} = X_3 + X_6 + X_9 = 20$$

The next step is to prepare the data for computer computation arranging the constraints as follows:

1. The less-than or equal inequalities

2. The strict equalities

3. The greater-than or equal inequalities

					Variables					
X_1	X_2	X_3	X_4	X_5	X_6	X_7	X_8	X_9		
1	1	1	0	0	0	0	0	0	$\leqslant$	24
0	0	0	0	0	1	0	0	0	$\leqslant$	4
0	0	0	1	1	1	0	0	0	$\leqslant$	30
0	0	0	0	0	0	1	1.4	1.8	$\leqslant$	25
1	0	0	1	0	0	1	0	0	$=$	21
0	1	0	0	1	0	0	1	0	$=$	27
0	0	1	0	0	1	0	0	1	$=$	20
0	0	0	0	0	0	0	1	0	$=$	0
0	0	0	1	1	1	0	0	0	$\geqslant$	26

We are now ready to utilize the time-share computer, in which we have stored a linear-programming program. The printout, Figure 1, shows the routines followed to determine the optimum costs of purchased parts for a given month.

Printout section A lists the entry of the coefficients of each of the variables from the preceding table of "Variables." B (line 19) enters the values in the right-hand column of the same table. C (line 20) shows the entry of the co-efficients of the objective function, or the costs for each size (from "Cost per part for engine size" table). Values are entered with negative signs to effect minimization.

FIGURE 1   Computer Printout for the Example Problem. See Text for Explanation.

A   10 DATA 1,1,1,0,0,0,0,0,0
11 DATA 0,0,0,0,0,1,0,0,0
12 DATA 0,0,0,1,1,1,0,0,0
13 DATA 0,0,0,0,0,0,1,1.4,1.8
14 DATA 1,0,0,1,0,0,1,0,0
15 DATA 0,1,0,0,1,0,0,1,0
16 DATA 0,0,1,0,0,1,0,0,1
17 DATA 0,0,0,0,0,0,0,1,0
18 DATA 0,0,0,1,1,1,0,0,0

B   19 DATA 24,4,30,25,21,27,20,0,26

C   20 DATA –647,–785,–842,–652,–762,–878,–683,–728,–893

D   L I N P R Ø

TYPE '2' FØR ØUTPUT ØF TABLEAUS AND BASIS AT EACH ITERATIØN,
    '1' FØR THE BASIS ØNLY, ØR
    '0' FØR JUST THE SØLUTIØN.
WHICH ?0

WHAT ARE M AND N ØF THE DATA MATRIX ?9,9

HØW MANY 'LESS THANS','EQUALS','GREATER THANS' ?4,4,1

YOUR VARIABLES   1 THROUGH 9
SURPLUS VARIABLES 10 THROUGH 10
SLACK VARIABLES 11 THROUGH 14
ARTIFICIAL VARIABLES 15 THROUGH 19

(continued)

E ANSWERS:

VARIABLE	VALUE	Supplier
1	4 ⬅——————— A	
4	3 ⬅——————— B	
10	4	
14	11	
7	14 ⬅——————— C	
12	4	
3	20 ⬅——————— A	
8	0	
5	27 ⬅——————— B	

ØBJECTIVE FUNCTIØN VALUE   −51520

DUAL VARIABLES:

CØLUMN	VALUE
10	0
11	36
12	0
13	31
14	0
15	−683
16	−793
17	−878
18	65.00001

After signaling the computer to run, the program asks for additional information, section D. The responding answers are underscored. The first response (0) schedules the amount of detail we would like to see concerning the solution of the problem. The second response (9,9) tells the program the size of the data matrix included (in this case) in lines 10 through 18. The third response (4,4,1) tells the program the equalities or inequalities of the data on line 19. For example, after this response, the program interprets line 19 to read $\leqslant 24, \leqslant 4, \leqslant 30, \leqslant 25, = 21, = 27, = 20, = 0, \geqslant 26$, and places each of these at the ends of lines 10 through 18.

Under "answers," section E, the values corresponding to variables 1 through 9 are taken from the printout. Identification of the suppliers has been written on the printout. As explained earlier, the variables X_n are assigned to the monthly quantities of a part for a specified number. The variable 1 in the printout corresponds to X_1 and its value is the number of 8-cylinder parts to be provided by Supplier A. The "objective function value,"–51520, is the optimum total cost of purchased parts for the month.

Other information in this printout, Figure 1, concerning surplus, slack, artificial, and dual variables concern intermediates set up by the program in arriving at a solution. This information may be disregarded in this case.

Table I shows the results of combining the values generated by the computer program with the earlier matrices.

TABLE I. COMPUTERIZED LINEAR PROGRAMMING DETERMINES OPTIMUM TOTAL COST OF PURCHASED PARTS

| Supplier | *Monthly costs of parts by engine sizes* | | | |
	8 Cylinder	*12 Cylinder*	*16 Cylinder*	*Totals*
A	$647 × 4 = $2,588		$842 × 20 = $16,840	$19,428
B	$652 × 3 = $1,956	$762 × 27 = $20,574		$22,530
C	$683 × 14 = $9,562			$ 9,562
Totals	$14,106	$20,574	$16,840	$51,520

We now have the order quantities for each part to be placed with each supplier to give us the minimum total cost for the order period. The program run cost approximately $1.00 of General Electric Time Sharing computer services.

When the technique was used for the first time in the Diesel Engine Products Department, the buyers were able to identify considerable savings over their previous best-choice combinations.

The event that triggered this undertaking was anticipated price increases by several suppliers simultaneously. The Materials organization resorted to linear programming to solve the problem. Once it was successfully solved, they realized they had a golden opportunity to use the tool for other components which were not originally considered problems, but were worth looking at to find improvements.

The major difficulty encountered was that of identifying and quantifying all of the constraints. This required input from several people, each knowledgeable in his own area. Their input was documented on flip charts during brainstorming sessions, and the charts were studied later for developing the constraints.

OPTIMUM PRESS
LOADINGS IN THE
BOOK MANUFACTURING
INDUSTRY*

Abstract: The purpose of this investigation is to determine how a given product mix should be loaded on feasible alternative press types so that total profits earned from printing are maximized, and press capacity is not exceeded. A linear programming model is formulated to derive a solution, utilizing the computer to facilitate the computational process.

problem statement

From sales plans and firm orders, a product mix can be identified at any given point in time. The ability to produce this work within a required time frame is limited to the available capacity on required equipment. When a press is "oversold", limited scheduling options are available. A job may be scheduled on a less economical press, the work can be farmed out to another printer, or the job can be lost to a competitor. Assuming farm outs are far more desirable than feeding the competition, and that available capacity for farming work out is virtually unlimited, the decision alternatives are narrowed to two: (1) loading alternative presses or (2) farming the work out. The scheduling problem then becomes one of loading available capacity (including farm out capacity) in such a way that profits are maximized and demand is satisfied. The general linear programming form is,

maximize,

$$\sum_{i=1}^{N} \sum_{j=1}^{N+1} p(i,j) \, x \, (i,j)$$

subject to,

$$\sum_{i=1}^{N} f(i,j) \, x \, (i,j) \leqslant B(j) \quad j = 1, \ldots ,N$$

*From a design project prepared for IE5900 at the University of Tennessee, Knoxville, by Charles H. Aikens III.

$$\sum_{j=1}^{N+1} x(i,j) = A(i) \qquad\qquad i = 1, \ldots ,N$$

$$x(i,j) \geqslant 0 \qquad\qquad\quad i = 1, \ldots \ldots , N$$
$$j = 1, \ldots \ldots , N+1$$

where,

N   = Number of press categories.

p(i,j) = Average profit per hour earned from work sold on press category (i) and scheduled on press category (j).

x(i,j) = Number of press hours sold on press category (i) and scheduled on press category (j).

f(i,j) = Factor which converts one hour on press category (i) to an equivalent number of hours on press category (j).

B(j)   = Capacity in hours of press category (j).

A(i)   = Demand for press category (i).

N+1   = This category represents farm out capacity.

the tableau

The problem as stated would imply an N by (N+1) matrix. However, each press is not a reasonable substitute for every other press. Press alternatives as they relate to each press category were identified and the nonfeasible (i,j) combinations are not considered in the problem solution. The A(i)'s for this problem are derived from the annual sales plans. The B(j)'s are calculated in Table I below. These values are based on a seven-day three-shift schedule, with every fourth Sunday down for shift rotation, a fifty-one week year, and 90% scheduling efficiency.

TABLE I   PRESS CAPACITY CHART

Days Available:	38 weeks @ 7 days	266 days
	13 weeks @ 6 days	78 days
	Total	344 days

(continued)

Category	Hours/Day	×	Days Available	×	No. Presses	×	Efficiency	Capacity
1	22.5		344		5		.9	34830
2	22.5		344		1		.9	6966
3	22.5		344		4		.9	30960
4	22.5		344		3		.9	20898
5	24.0		344		4		.9	29721
6	24.0		344		1		.9	7430
7	24.0		344		2		.9	14860

The resultant tableau is shown in Table II. An empty cell is indicated for each cell where the (i,j) combination is non-feasible.

TABLE II   TABLEAU

Press Sold (i)	Press Scheduled (j)								↓A(i)
	1	*2*	*3*	*4*	*5*	*6*	*7*	*8*	
1	x(1,1)	—	x(1,3)	—	—	—	—	x(1,8)	55555
2	x(2,1)	x(2,2)	x(2,3)	—	—	—	—	x(2,8)	1667
3	—	—	x(3,3)	x(3,4)	—	—	—	x(3,8)	27228
4	—	—	x(4,3)	x(4,4)	—	—	—	x(4,8)	10000
5	x(5,1)	—	x(5,3)	—	x(5,5)	—	—	x(5,8)	22222
6	x(6,1)	x(6,2)	—	—	x(6,5)	x(6,6)	x(6,7)	x(6,8)	12500
7	x(7,1)	—	x(7,3)	—	—	—	x(7,7)	x(7,8)	18000
B(j)	34830	6966	30960	20898	29721	7430	14860	∞	

**press hour
conversion factors**

The formulation of one set of constraint equations depends on the calculation of the press hour conversion factors, the f(i,j)'s. The factors are constants. Each f(i,j) is defined as,

$$f(i,j) = \frac{I(i)}{I(j)} \times \frac{sp(i)}{sp(j)} \qquad \begin{array}{l} i = 1, \ldots 7 \\ j = 1, \ldots 7 \end{array}$$

where

$I(i)$ =   most common imposition for press cateogry (i)

$sp(i)$ =   net impressions per hour for press category (i)

All $I(i)$ and $sp(i)$ values are summarized in Table III.

TABLE III   MOST COMMON IMPOSITIONS AND NET PRESS SPEEDS

Press Category (i)	I(i) Most Common Imposition	sp(i) Net Impressions Per Hour
1	128 pages	4000
2	64 pages	4000
3	64 pages (one side)	4000
4	64 pages (one side)	4000
5	32 pages	16000
6*	not applicable	1789
7	64 pages	12000

Conversion factors are summarized in Table IV.

TABLE IV   PRESS HOUR CONVERSION MATRIX f(i,j)

Press Sold	Press Scheduled						
	1	2	3	4	5	6	7
1	1	–	2	–	–	–	–
2	.5	1	1	–	–	–	–
3	–	–	1	1	–	–	–
4	–	–	2	1	–	–	–
5	1	–	2	–	1	–	–
6	2.6	5.2	–	–	2.6	1	1.74
7	1.5	–	3	–	–	–	1

*Press category 6 is a complete book system and the concept of impositions does not apply. Conversion factors were computed as the number of hours required to produce an average book on each alternative press category relative to press category 6.

the reduced form

Considering only the feasible (i,j) combinations, the problem is reduced from a field of 56 variables (7 × 8) to 27 variables. Renaming each x(i,j) to an x(k) as shown in Table V, the linear programming model becomes,

maximize

$$Z = \sum_{k=1}^{27} p(k)\, x(k)$$

or minimize

$$-Z$$

subject to,

x(1) + x(2) + x(3)	= 55555
x(4) + x(5) + x(6) + x(7)	= 1667
x(8) + x(9) + x(10)	= 27228
x(11) + x(12) + x(13)	= 10000
x(14) + x(15) + x(16) + x(17)	= 22222
x(18) + x(19) + x(20) + x(21) + x(22) + x(23)	= 12500
x(24) + x(25) + x(26) + x(27)	= 18000
x(1) + .5(x(4)) + x(14) + 2.6(x(18)) + 1.5(x(24))	⩽ 34830
x(5) + 5.2(x(19))	⩽ 6966
2(x(2)) + x(6) + x(8) + 2(x(11)) + 2(x(15)) + 3(x(25))	⩽ 30960
x(9) + x(12)	⩽ 20898
x(16) + 2.6(x(20))	⩽ 29721
x(21)	⩽ 7430
1.74(x(22)) + x(26)	⩽ 14860
x(k) ⩾ 0 k = 1,. . . . 27	

TABLE V   RENAMING VARIABLES

x(i,j)	x(k)	x(5,1)	x(14)
x(1,1)	x(1)	x(5,3)	x(15)
x(1,3)	x(2)	x(5,5)	x(16)
x(1,8)	x(3)	x(5,8)	x(17)
x(2,1)	x(4)	x(6,1)	x(18)
x(2,2)	x(5)	x(6,2)	x(19)
x(2,3)	x(6)	x(6,5)	x(20)
x(2,8)	x(7)	x(6,6)	x(21)
x(3,3)	x(8)	x(6,7)	x(22)
x(3,4)	x(9)	x(6,8)	x(23)
x(3,8)	x(10)	x(7,1)	x(24)
x(4,3)	x(11)	x(7,3)	x(25)
x(4,4)	x(12)	x(7,7)	x(26)
x(4,8)	x(13)	x(7,8)	x(27)

assumptions

Since product mix is defined relative to the number of hours sold on a particular press category, it is necessary to make several simplifying assumptions concerning the specific products that will be manufactured. Equipment selection rules and the determination of viable alternate equipment types for any operation depend on such specifications as trim size, paper basis weight, number of pages in the book, bulk of the book, number of colors per page, etc. Through careful research an attempt has been made to duplicate average conditions. The average conditions found are detailed by means of the below listed assumptions.

1. Any job sold on press category (i) may be run on any press category (j) identified as a feasible alternative.

2. Binding capacity is adequate to produce any work that is printed.

3. If demand for a given press category exceeds the available capacity of that press category added to the available capacity of all feasible alternatives for that category, the excess capacity can be farmed out to another printer.

4. Equipment manning is flexible enough to work any number of hours up to the maximum.

5. The average edition quantity for all production jobs is 15,000.

6. The average book contains twelve signatures including two endpapers.

7. The average number of pages per book for each binding style is as shown in Table VI.

8. The most common trim size for books produced on the belt press/binder system is 5½" x 8½".

TABLE VI   AVERAGE NUMBER OF PAGES PER BOOK

Binding Style	*Average Number of Pages*
Smyth Sewn	452
Saddle Stitched	104
Side Stitched	356
Adhesive Bound, Hard Cover	368
Adhesive Bound, Paper Cover	216

**optimal production
schedule**

The problem consists of twenty-seven real variables, seven slack variables, and seven artificial variables, or forty-one tableau columns; and fourteen constraints, or tableau rows. A problem this large is rather cumbersome to solve using manual techniques. Therefore, a computer program was used to obtain the optimum solution. The profit matrix, $p(i,j)$, shown in Table VII was developed and used in determining the solution.

After twenty-one iterations, the optimum value of the objective function was calculated to be $Z = \$10,189,780$ profit. The individual values of $x(k)$ are summarized in Table VIII.

This solution reveals some interesting points. All press categories, except press category 4, should be loaded to capacity. Press category 4 should be loaded only to satisfy demand for that press type, or 48% capacity. This means press category 4 should only be scheduled on a five-day, two-shift basis. The optimal solution is to farm out 48% of the demand for press category 1 while scheduling work "oversold" on categories 6 and 7 on category 1. It is also significant to note that it is more profitable to schedule press category 1 work on press category 3 at double the press time than to schedule that work on category 1 and produce categories 6 and 7 work elsewhere or farm that work out. For each press category "undersold" (categories 2, 3, 4 and 5), it is most profitable to schedule the entire demand on that category. This result would be predictable.

TABLE VII THE PROFIT MATRIX
p(i,j) COEFFICIENTS

				Press Scheduled				
Press Sold	1	2	3	4	5	6	7	8
1	113	–	59	–	–	–	–	44
2	87	80	60	–	–	–	–	45
3	–	–	73	38	–	–	–	29
4	–	–	-133	74	–	–	–	-166
5	- 27	–	- 82	–	114	–	–	-103
6	-691	-733	–	–	-339	167	-256	-916
7	-126	–	-208	–	–	–	144	-260

TABLE VIII SOLUTION TABLEAU

				Press Scheduled					
Press Sold	1	2	3	4	5	6	7	8	
1	27086.5	–	1866	–	–	–	–	26602.5	55555
2	0	1667	0	–	–	–	–	0	1667
3	–	–	27228	0	–	–	–	0	27228
4	–	–	0	10000	–	–	–	0	10000
5	0	–	0	–	22222	–	–	0	22222
6	1166.7	1019.1	–	–	2884.2	7430	0	0	12500
7	3140	–	0	–	–	–	14860	0	18000
	34830	6966	30960	10000	29721	7430	14860		

application of results

The validity of the problem solution depends on the accuracy of the data entering the model. Having established the model and technique, the necessary data can be purified over time. For example, most common impositions or average page counts could be improved through detailed historical analysis.

For the purposes of this study, the press demand used was the submitted sales plan for FY75, and the time period was one year. This provided meaningful results for use in strategic planning. However, perhaps a better application for the technique is the use of the specific demand experienced in a shorter period of time, e.g. 90 days.

The optimum solution under those conditions would aid in making such scheduling decisions as 1) what work should be farmed out, 2) selection of the "best" alternative presses when schedules are unacceptable to customers, and 3) when specific types of equipment should be placed on extended schedules.

INTRODUCTION TO

LINEAR

PROGRAMMING FOR

PRODUCTION*

Abstract: The basic features of linear programming as applied to a production problem are illustrated in this study. Formulation of the problem, collection of data and presentation of results are carefully described by the author.

Enormously complicated industrial situations can be optimized with the computer, using the technique of linear programming. But because of the complexity, linear programming is rarely successful unless key people from each of the related functional areas are involved in setting out the problem.

One of these key people is very often the industrial engineer. Even if you are not mathematically-oriented, it pays to understand the principles of LP so that you at least understand what the mathematicians are trying to do to your plant.

In the future, it is expected that LP models will be used even more extensively, not only for the tactical decisions of production and distribution, but also for the strategic decisions of marketing, long-range planning, diversification, and acquisition. As a first step toward understanding LP, consider a simplified LP production model formulated for a hypothetical company.

This company has two plants producing a total of four products according to the flow diagrams in Figure 1. In each plant, the raw material (ARM and BRM) is first processed into an intermediate product. The intermediate product is further processed and then goes either into the production of edible oils (ALIP

*By J. Frank Sharp. Reprinted by permission of the publisher from *The Journal of Industrial Engineering*, December 1970, pp. 10-16.

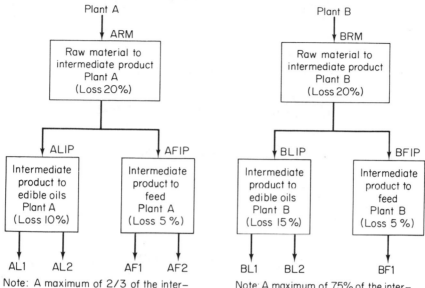

Plant A

↓ ARM

Raw material to
intermediate product
Plant A
(Loss 20%)

↓ ALIP   ↓ AFIP

Intermediate   Intermediate
product to   product to
edible oils   feed
Plant A   Plant A
(Loss 10%)   (Loss 5 %)

AL1   AL2   AF1   AF2

Note: A maximum of 2/3 of the intermediate product at Plant A can go to the
production of edible oils.

Plant B

↓ BRM

Raw material to
intermediate product
Plant B
(Loss 20%)

↓ BLIP   ↓ BFIP

Intermediate   Intermediate
product to   product to
edible oils   feed
Plant B   Plant B
(Loss 15%)   (Loss 5 %)

BL1   BL2   BF1

Note: A maximum of 75% of the intermediate product at Plant B can go to the
production of edible oils.

FIGURE 1   Flow Diagram Shows the Production Constraints on the Raw
Material, Intermediate Material, and Finished Products

and BLIP) or feed (AFIP and BFIP). Plant A can produce two oil products (AL1
and AL2) and two feed products (AF1 and AF2). Plant B can produce the same
two oil products (BL1 and BL2), but only the first feed product (BF1).

The loss values in Figure 1 indicate the production yield at various points
in the production process. At plant A, 100 pounds of raw material (ARM) is
necessary to produce 80 pounds of intermediate product (ALIP or AFIP), while
100 pounds of intermediate product (ALIP) is necessary to produce 90 pounds
of oil product, etc.

Additional information needed for the analysis is given in Tables I through
IV. This includes capacity restrictions on the amount of raw material that each
plant can process, the amount of intermediate product that can be processed,
the amount of each product that can be produced, the forecasted demand and
price for each product, the processing costs for each production process, and the
raw material costs.

Given the prices, production costs, raw material costs, and production
relationships, the unit contribution to profits of each product at each plant can
be determined. For instance, AF1's contribution per 100 pounds of final

product would be:

$12.00 (Price)

− 1.10 (Final stage production costs)

$$- \frac{1.00}{(0.9)} \Bigg\} \text{(Intermediate stage production costs)}$$

$$- \frac{0.50}{(0.8)\,(0.9)} \Bigg\} \text{(First-stage production costs)}$$

$$- \frac{2.50}{(0.8)\,(0.9)} \Bigg\} \text{(Raw material costs)}$$

$$= \$5.662$$

All of the calculated unit contributions are given in Table V.

TABLE I.   PROCESSING CAPACITIES (100 POUNDS/PERIOD)

Plant A		Plant B		Description
KARM	4000	KBRM	3000	Raw material processing capacity.
KALIP	2500	KBLIP	1700	Processing capacity of intermediate product going to edible oils.
KAFIP	1250	KBFIP	1100	Processing capacity of intermediate product going to feeds.
KAL1	1500	KBL1	1000	Processing capacity for edible oil product 1.
KAL2	1500	KBL2	1000	Processing capacity for edible oil product 2.
KAF1	750	KBF1	1000	Processing capacity for feed product 1.
KAF2	750	—		Processing capacity for feed product 2.

TABLE II. FORECASTED DEMANDS
(100 POUNDS/PERIOD)
AND PRICES ($/100 POUNDS)

Demand		Price	Product
DML1	2000	$12	Edible oil product 1
DML2	2000	$15	Edible oil product 2
DMF1	1200	$ 8	Feed product 1
DMF2	500	$ 7	Feed product 2

TABLE III. PROCESSING COSTS FOR EACH
PRODUCTION PROCESS ($/100 POUNDS)

Process	Plant A	Plant B
L1	$1.10	$1.20
L2	1.20	1.30
F1	0.90	0.80
F2	0.70	X
LIP	1.00	1.00
FIP	0.95	0.60
RM	0.50	0.60

TABLE IV. RAW MATERIAL
COSTS ($/100 POUNDS)

Plant A	$2.50
Plant B	2.60

TABLE V.   CONTRIBUTION FOR
EACH PRODUCT ($/100 POUNDS)

Product	Plant A	Plant B
L1	$5.662	$4.918
L2	8.552	7.818
F1	2.153	2.358
F2	1.353	X

**production facts put
in LP form**

The first step towards optimizing this plant's production is to arrange all the known factors, or constraints, in the form of matrix, Figure 2. The matrix is like a set of simultaneous equations, except that some of the equations are actually inequalities. There are an infinite number of solutions to the expressions in the matrix. The aim is to find the solution that will optimize a stated function, such as profit.

In the matrix of Figure 2, the names of products and intermediate products are given across the top. These are the variables of the equations and inequalities that make up the matrix. Whenever one of these appears in an equation or inequality its coefficient is written directly below it.

The first production constraint for each plant (Lines MAIL and MAIF) represents the material balance between the amount of edible oil products produced and the amount of intermediate products required. Similarly, the second production constraint for each plant represents the material balance between the amount of feed products produced and the amount of intermediate products needed. The data for these constraints comes from Figure 1. For instance:

$$AL1 + AL2 = 0.9 \, ALIP$$

due to the 10 percent production loss in this process. This is equivalent to the first production constraint

$$AL1 + AL2 - 0.9 \, ALIP = 0$$

The third production constraint for each plant (line MARI) represents the material balance between the amount of intermediate product produced and the amount of raw material required. For instance, from Figure 1,

	AL1	AL2	AF1	AF2	ALIP	AFIP	ARM	BL1	BL2	BF1	BLIP	BFIP	BRM		
REV	12.00	15.00	8.00	7.00				12.00	15.00	8.00					
CØST	1.10	1.20	0.90	0.70	1.00	0.95	3.00	1.20	1.30	0.80	1.00	0.60	3.20		
CØNTR	5.622	8.522	2.153	1.353				4.918	7.818	2.358					
MAIL	1.	1.			-0.9									=	0.
MAIF			1.	1.		-0.95								=	0.
MARI					1.	1.	-0.8							=	0.
RALF					1.		-2.							≤	0.
KAL1	1.													≤	1500.
KAL2		1.												≤	1500.
KAF1			1.											≤	750.
KAF2				1.										≤	750.
KALIP					1.									≤	2500.
KAFIP						1.								≤	1250.
KARM							1.							≤	4000.
MBIL								1.	1.		-0.85			=	0.
MBIF										1.		-0.95		=	0.
MBRI											1.	1.	-0.8	=	0.
RBLF											1.		-3.	≤	0.
KBL1								1.						≤	1000.
KBL2									1.					≤	1000.
KBF1										1.				≤	1000.
KBLIP											1.			≤	1700.
KBFIP												1.		≤	1100.
KBRM													1.	≤	3000.
DL1	1.							1.						≤	2000.
DL2		1.							1.					≤	2000.
DF1			1.							1.				≤	1200.
DF2				1.										≤	500.

AL1, AL2, AF1, AF2, ALIP, AFIP, ARM, BL1, BL2, BF1, BLIP, BFIP, BRM, ≥ 0.

FIGURE 2   The Linear Programming Model Is Really a Set of Equations and Inequalities That Expresses the Production Conditions of the Plant

$$\text{ALIP} + \text{AFIP} = 0.8\ \text{ARM}$$

This constraint indicates that the amount of intermediate product produced at plant A is 0.8 of the raw material processed, due to the 20 percent production loss in this process. In the matrix, all of the variables are moved to the left-hand side of the equality and placed in their respective columns. Also from Figure 1,

$$\text{ALIP} \leqslant (2/3)\,(0.8)\ \text{ARM}$$

because of the limitation on the relative production of ALIP. But:

$$0.8\ \text{ARM} = \text{ALIP} + \text{AFIP}$$

Combining these two expressions produces the inequality:

$$\text{ALIP} - 2\text{AFIP} \leqslant 0$$

which is the fourth production constraint for plant A (Line RALF).

In addition, there are production constraints that give the maximum production rate for each process. These are based on the data in Table I. For instance, $\text{AL1} \leqslant 1500$ represents the processing capacity for edible oil product one during the period (Line KALI).

The first three rows of the matrix are merely a listing of revenue, cost, and contribution functions for each product. The next 11 rows represent production constraints at plant A, the next ten rows represent production constraints at plant B, while the next four rows represent the forecasted demand for each of the four products. The last row merely states the requirement that no variable can take on negative values.

The CONTR function is the one actually to be optimized. This expression gives expected total contribution of all the products as listed in Table V—max $\text{CONTR} = \$5{,}622\text{AL1} + 8.522\ \text{AL2} + 2.153\text{AF1} + 1.353\text{AF2} + 4.918\ \text{BL1} + 7.818\text{BL2} + 2.358\text{BF1}$. The REV function gives the corresponding total revenue while the COST function gives the corresponding total costs. The coefficients for the REV function are the prices in Table II. The coefficients for the COST function are the processing costs from Table III and the raw material costs from Table IV. The value of the REV function minus the value of the COST function equals the value of the CONTR function.

There is one demand constraint for each of the four products. It is assumed that for each product the firm can sell as much as it can produce, up to the forecasted demand. For this example, production capacity is insufficient to satisfy all demands. The constraint for edible oil product one is

$$\text{AL1} + \text{BL1} \leqslant 2000. \text{ (Line DL1)}$$

The forecasted demands are from Table II.

Even if production capacity was sufficient to meet all demands, the same sort of constraints could be used. Then all profitable demands would be satisfied.

now find the solution

Even this simplified example would require several days of hand calculations in order to obtain an optimal solution. However an optimal solution can be obtained in much less time, at much less expense, with much less chance for error, using any one of the several canned LP computer codes. This particular example was solved on an IBM Model 360/50 using IBM's MPS/360. The computer time (CPU) was less than 15 seconds for the 15 iterations necessary to obtain an optimal solution.

This example has 28 rows and 13 columns (not counting slack columns). A real-world problem might have several hundred rows and over a thousand columns. Instead of a few seconds, the computer time needed might be an hour or more. (This time can often be reduced considerably by using some of the special options available in the more sophisticated computer codes. For instance, if the first run has been completed and a few modifications made, the computer can start the second run with the previous optimal solution instead of starting from scratch.)

The computer input data is shown in Figure 3 (as output by the computer code). The objective function is coded as OBJ instead of CONTR. A portion of the computer output is shown in Figures 4, 5, and 6. Figure 4 gives the optimal value of the objective function as $27,068.

Figure 5 lists the optimal value of the left-hand side of each function and constraint under ACTIVITY. This plus the optimal SLACK ACTIVITY is equal to the RHS value for each constraint. Note that all are under UPPER LIMIT in this example. All of these values in rows 4 through 28 are in 100's of pounds per period. Also given in Figure 5 is the optimal value of the DUAL ACTIVITY (sometimes called shadow price or imputed value). The DUAL ACTIVITY gives the changes in the objective function if the RHS for that constraint is decreased by one unit. The negative of this value gives the change in the objective function if the RHS for that constraint is increased by one unit. The units are dollars per 100 pounds.

Figure 6 gives the optimal activity level for each variable (under ACTIVITY). The units are 100's of pounds per period.

distribution costs
add complexity

Some idea of the complexity of real-life problems can now be realized by considering the fact that product costs to the customer depend on transportation as well as production costs. It would be perfectly possible to take the production levels found by the above LP model and find the shipment schedule that would minimize transportation costs. However, the result would only be suboptimum. To find the overall optimum, the LP model must be modified to include both production and transportation.

Instead of one activity representing the production of each product at each plant, there would be one activity representing each customer with a demand for that product. For instance, if there are three customers for the first

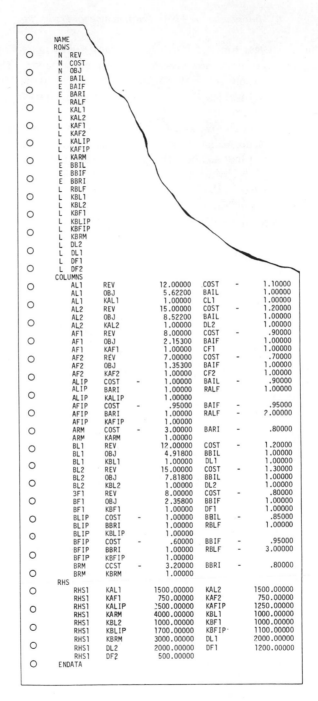

```
NAME
ROWS
 N   REV
 N   COST
 N   OBJ
 E   BAIL
 E   BAIF
 E   BARI
 L   RALF
 L   KAL1
 L   KAL2
 L   KAF1
 L   KAF2
 L   KALIP
 L   KAFIP
 L   KARM
 E   BBIL
 E   BBIF
 E   BBRI
 L   RBLF
 L   KBL1
 L   KBL2
 L   KBF1
 L   KBLIP
 L   KBFIP
 L   KBRM
 L   DL2
 L   DL1
 L   DF1
 L   DF2
COLUMNS
    AL1    REV        12.00000   COST   -    1.10000
    AL1    OBJ         5.62200   BAIL        1.00000
    AL1    KAL1        1.00000   CL1         1.00000
    AL2    REV        15.00000   COST   -    1.20000
    AL2    OBJ         8.52200   BAIL        1.00000
    AL2    KAL2        1.00000   DL2         1.00000
    AF1    REV         8.00000   COST   -     .90000
    AF1    OBJ         2.15300   BAIF        1.00000
    AF1    KAF1        1.00000   CF1         1.00000
    AF2    REV         7.00000   COST   -     .70000
    AF2    OBJ         1.35300   BAIF        1.00000
    AF2    KAF2        1.00000   CF2         1.00000
    ALIP   COST   -    1.00000   BAIL   -     .90000
    ALIP   BARI        1.00000   RALF        1.00000
    ALIP   KALIP       1.00000
    AFIP   COST   -     .95000   BAIF   -     .95000
    AFIP   BARI        1.00000   RALF   -    ?.00000
    AFIP   KAFIP       1.00000
    ARM    COST   -    3.00000   BARI   -     .80000
    ARM    KARM        1.00000
    BL1    REV        12.00000   COST   -    1.20000
    BL1    OBJ         4.91800   BBIL        1.00000
    BL1    KBL1        1.00000   DL1         1.00000
    BL2    REV        15.00000   COST   -    1.30000
    BL2    OBJ         7.81800   BBIL        1.00000
    BL2    KBL2        1.00000   DL2         1.00000
    3F1    REV         8.00000   COST   -     .80000
    BF1    OBJ         2.35800   BBIF        1.00000
    BF1    KBF1        1.00000   DF1         1.00000
    BLIP   COST   -    1.00000   BBIL   -     .85000
    BLIP   BBRI        1.00000   RBLF        1.00000
    BLIP   KBLIP       1.00000
    BFIP   COST   -     .60000   BBIF   -     .95000
    BFIP   BBRI        1.00000   RBLF   -    3.00000
    BFIP   KBFIP       1.00000
    BRM    CCST   -    3.20000   BBRI   -     .80000
    BRM    KBRM        1.00000
RHS
    RHS1   KAL1     1500.00000   KAL2     1500.00000
    RHS1   KAF1      750.00000   KAF2      750.00000
    RHS1   KALIP    2500.00000   KAFIP    1250.00000
    RHS1   KARM     4000.00000   KBL1     1000.00000
    RHS1   KBL2     1000.00000   KBF1     1000.00000
    RHS1   KBLIP    1700.00000   KBFIP    1100.00000
    RHS1   KBRM     3000.00000   DL1      2000.00000
    RHS1   DL2      2000.00000   DF1      1200.00000
    RHS1   DF2       500.00000
ENDATA
```

FIGURE 3   Names of All Rows and Columns and the Values of the Coefficients That Make Up the Matrix Are First Input to the Computer

```
SECTION 1 - ROWS
```

NUMBER	...ROW..	AT	...ACTIVITY...	SLACK ACTIVITY	..LOWER LIMIT.	..UPPER LIMIT.	.DUAL ACTIVITY
1	REV	BS	59328.33333	59328.33333-	NONE	NONE	.
2	COST	BS	32261.00000-	32261.00000	NONE	NONE	.
3	OBJ	BS	27067.86000	27067.86000-	NONE	NONE	1.00000
4	BAIL	EQ	.	.	.	.	5.62200-
5	BAIF	EQ	.	.	.	.	1.35300-
6	BARI	EQ	.	.	.	.	3.80165-
7	RALF	UL	.	.	NONE	.	1.25815-
8	KAL1	BS	420.00000	1080.00000	NONE	1500.00000	.
9	KAL2	UL	1500.00000	.	NONE	1500.00000	.
10	KAF1	BS	535.00000	215.00000	NONE	750.00000	.
11	KAF2	BS	478.33333	271.66667	NONE	750.00000	.
12	KALIP	BS	2133.33333	366.66667	NONE	2500.00000	.
13	KAFIP	BS	1066.66667	183.33333	NONE	1250.00000	.
14	KARM	LL	4000.00000	.	NONE	4000.00000	3.04132-
15	BBIL	EQ	.	.	.	.	4.91800-
16	BBIF	EQ	.	.	.	.	1.55800-
17	BBRI	EQ	.	.	.	.	1.48010-
18	RBLF	BS	400.00000-	400.00000	NONE	.	.
19	KBL1	BS	945.00000	55.00000	NONE	1000.00000	.
20	KBL2	BS	500.00000	500.00000	NONE	1000.00000	.
21	KBFI	BS	665.00000	335.00000	NONE	1000.00000	.
22	KBLIP	LL	1700.00000	.	NONE	1700.00000	2.70020-
23	KBFIP	BS	700.00000	400.00000	NONE	1100.00000	.
24	KBRM	UL	3000.00000	.	NONE	3000.00000	1.18408-
25	DL1	BS	1365.00000	635.00000	NONE	2000.00000	.
26	DL2	UL	2000.00000	.	NONE	2000.00000	2.900000-
27	DF1	LL	1200.00000	.	NONE	1200.00000	.800000-
28	DF2	BS	478.33333	21.66667	NONE	500.00000	.

FIGURE 4 Listed under ACTIVITY Are the Optimum Values for Each Row; DUAL ACTIVITY Shows the Sensitivity of the Objective Function to Each Constraint

oil product, then instead of one variable for plant A (AL1) and one variable for plant B (BL1) there would be three variables for each plant (AL11, AL12, AL13, BL11, BL12, and BL13). The coefficients of AL11, AL12, and AL13 in the objective function would differ from the previous coefficient of AL1 only by the corresponding unit transportation cost. (This will vary, depending on who pays for transportation.) Instead of one demand constraint for this product, there would now be one for each customer. For instance, the demand constraint representing the first customer for the first oil product would be

$$AL11 + BL11 \leq DML11$$

Some of the possible outputs of an LP production-distribution model are the expected values of:

Maximum contribution

Total cost and revenue breakdown

Production levels for each product

Raw materials requirement

Shipments schedule

Process usage

Unfilled demands

Shadow prices for each process

Shadow prices for additional demands

For this example, these items can be found in the basic computer output shown in Figures 4 and 5, but for a large model the output can be voluminous. Also much of the output, such as row and activity names, will be in coded form and will be unintelligible until decoded. The computer output will usually give values to several decimal places. It is usually ridiculous to assume such accuracy, and values should be rounded off.

```
SECTION 2 - COLUMNS

NUMBER  .COLUMN.  AT  ...ACTIVITY...  ..INPUT COST..

    29  AL1       BS       420.00000       5.62200
    30  AL2       BS      1500.00000       8.52200
    31  AF1       BS       535.00000       2.15300
    32  AF2       BS       478.33333       1.35300
    33  ALIP      BS      2133.33333       .
    34  AFIP      BS      1066.66667       .
    35  ARM       BS      4000.00000       .
    36  BL1       BS       945.00000       4.91800
    37  BL2       BS       500.00000       7.81800
    38  BF1       BS       665.00000       2.35800
    39  BLIP      BS      1700.00000       .
    40  BFIP      BS       700.00000       .
    41  BRM       BS      3000.00000       .
```

FIGURE 5 ACTIVITY Listing Shows the Optimum Activity Level for Each Variable

In addition, certain members of management may be interested in only a small part of the output. For this purpose it is usually necessary to prepare a management report or reports. A FORTRAN program can be written so that the reports will be automatically generated, given the output of the LP computer code as input.

Figure 6 is a possible management report for the LP model of Figure 2. A shipment schedule is not given since distribution was not considered in this model. Due to rounding off, there may be slight variations from the original computer output.

Some of the practical problems involved in the formulation and implementation of a product-distribution model are:

Production. What level of detail should be used in modeling the plants? What processes should be included? Should products be grouped according to process?

Distribution. Should customers be lumped together into regions? What are appropriate regions?

Accounting. Are production costs broken down according to any rational system? How can the appropriate costs be obtained? How should the raw material be costed? How should by-products be costed?

Marketing. How can reasonable demand forecasts be obtained? Should products be grouped according to function, type of customer, etc? What processes should be used?

Market research. Are changes in the product line foreseen?

Finance. What are the costs of changes in production facilities? Are new facilities proposed?

Data processing systems. Is the necessary input data available in a readily available form? What systems work will be required?

Computer facilities. What in-house facilities are available? What outside facilities are available?

Personnel. What personnel are available? What is their experience? What are the contacts and availability of help in the above functional areas? Is the proposed completion date reasonable?

A linear programming model can be useful in aiding both tactical and strategic decision making. However, because it is a model and the output is only an approximation, it should be used only as a guide. Knowledge of both the possible uses and limitations of the model is also necessary. One of the greatest benefits of formulating and implementing a model is often the insight gained into the operations of the firm.

Make the model as compact as possible. Try to group products according to similar processing requirements or similar marketing characteristics, also try to group customers. Avoid unimportant constraints. This is extremely important if the model is ever going to be implemented.

Formulating a model requires a long-term commitment of resources. Estimates must be realistic. Personnel in several functional areas should be assigned, if only part-time. Even after the model is completed, personnel will be necessary for updating, regular runs, and strategic analyses.

REFERENCES

(1) Alexander, Thomas "Computers Can't Solve Everything," *Fortune*, LXXX, Number 5, October 1969.

(2) CEIR *Conference on Mathematical Model Building in Economics and Industry*, Hafner, 1968.

(3) Greene, James H., *Production Control: Systems and Decision*, Richard D. Irwin, Inc., Homewood, Illinois, 1965.

(4) Hadley, George, *Linear Programming*, Addison-Wesley Publishing Company, Reading, Massachusetts, 1962.

(5) Hess, Sidney, "Operations Research in the Chemical and Pharmaceutical Industries," R. T. Eddison and D. B. Hertz (editors). *Progress in Operations Research*, John Wiley and Sons, Inc., New York, 1964.

(6) Hertz, David, *New Power for Management Computer Systems and Management Science*, McGraw-Hill Book Company, New York, 1967.

(7) Sears, G. W., "Petroleum," R. T. Eddison and D. B. Hertz (editors), *Progress in Operations Research*, John Wiley and Sons, Inc., New York, 1964.

(8) Withington, Frederic, *The Real Computer: Its Influence, Uses, and Effects*, McGraw-Hill Book Company, New York, 1969.

FIGURE 6   A Typical Management Report

A.   Maximum Contribution ($/Period)

$59,326. Revenue
−32,261. Costs
$27,065. Maximum Contribution

B.   Sales Volume (100 pounds/period) and Revenue ($/period) Report

Product	Expected Sales	Unit Price	Revenue	Forecasted Demand	Unfulfilled Demand	Shadow Price
L1	1365	$12.00	$16,380	2000	635	0.00
L2	2000	15.00	30,000	2000	0	2.90
F1	1200	8.00	9,600	1200	0	0.80
F2	478	7.00	3,346	500	22	0.00
			$59,326			

C.   Processing Usage (100 pounds/period) and Cost ($/period) Report

Process	Expected Usage	Capacity	Processing Cost	Processing Cost	Excess Capacity	Shadow Price
AL1	420	1500	$1.10	$ 462.	1080	0
AL2	1500	1500	1.20	1,800.	0	0
AF1	535	750	.90	481.	215	0
AF2	478	750	.70	335.	272	0
ALIP	2133	2500	1.00	2,133.	366	0
AFIP	1067	1250	.95	1,014.	183	0
ARM	4000	4000	.50	2,000.	0	3.04
				$ 8,225.		
BL1	945	1000	1.20	1,134.	55	0
BL2	500	1000	1.30	650.	500	0
BF1	665	1000	.80	532.	335	0
BLIP	1700	1700	1.00	1,700.	0	2.70
BFIP	700	1100	.60	420.	400	0
BRM	3000	3000	.60	1,800.	0	1.18
				$ 6,236.		
				$14,461.		

D.   Raw Material Requirements (100 pounds/period) and Cost ($/period) Report

Plant	Requirements	Unit Cost	Raw Material Cost
A	4000	2.50	$10,000
B	3000	2.60	7,800
			$17,800

E.   Production Report (100's pounds/period)

Product	Plant A	Plant B	Total
L1	420.	945.	1365.
L2	1500.	500.	2000.
F1	535.	665.	1200.
F2	478.	X	478.

LINEAR PROGRAMMING— WITHOUT THE MATH*

Abstract: These two studies demonstrate the formulation of linear programming problems. Approximate solutions are determined by a trial and error process that will solve many industrial problems. However, sophisticated methods would probably be needed to solve more complex problems.

find best stocking plan

Stocking of finished goods when the shipping pattern shows a substantial seasonal variation is a problem that can be greatly simplified by constructing a matrix. Consider a consumer-specialty appliance manufacturer whose shipping schedule is low for the first eight months of the year, and peaks at three to four times that level during the last three months—far above capacity. Monthly capacity is 100 units on straight time and 20 units on overtime. The shipping demands are given in Table I.

TABLE I.   SHIPPING DEMANDS ON APPLIANCE MANUFACTURER PEAK AT NEARLY DOUBLE ITS CAPACITY DURING LAST MONTHS OF YEAR

Month:	6	7	8	9	10	11	12	Total
Units	60	60	60	90	140	200	220	830

*By Edward Cochran. Reprinted by permission of the publisher from *The Journal of Industrial Engineering*, November 1970, pp. 14-23.

The variable unit cost (labor, material, and variable shop overhead) is $100 at straight time, and $110 under overtime conditions. If a unit is held in inventory, a carrying charge of $5 per month is added to the straight-time cost. Units produced by overtime are shipped the same month, since their number is relatively small.

The problem is to find out how many units can be economically manufactured in low-demand months and held in inventory until needed. To analyze this problem, consider that there are two sources of production each month: straight time and overtime, with capacities of 100 and 20 units per month and costs of $100 and $110 per unit, respectively. For each month the shipping needs can be produced in any month preceding it, as well as in the month itself, provided that carrying charges are covered.

A matrix can be set up as shown in Figure 1. The figures in the corner of each cell show the cost difference over the minimum cost of $100 per unit. The bold figures show one low-cost production plan. This plan is constructed by starting with the lower right-hand corner and dealing first with high-demand months. This plan is easy to find because of the large number of cells in which entries are impossible. For instance, units cannot be shipped before they are built; no units are produced on overtime for inventory. There are several other plans of equal cost.

The final plan is summarized in Table II. Note that even a month of low shipments may have to be supplied out of inventory in order to permit later high-volume requirements to be met economically.

The total cost of the shipments, calculated as the sum of production cost and carrying charges, is $88,600 or $106.75 per unit.

Note that total production for the last seven months is only 760 units—70 units less than the 830 units shipped. This means that an inventory of 70 units must be provided from May (and perhaps April) production. It is usually necessary to analyze several months on each side of the period under consideration to ensure a suitable decision.

TABLE II. THE FINAL PLAN: NUMBER OF UNITS SHIPPED EACH MONTH OUT OF STRAIGHT-TIME PRODUCTION OVERTIME, AND INVENTORY

Month:	6	7	8	9	10	11	12
Shipments	60	60	60	90	140	200	220
Sources							
STP	—	—	—	—	—	—	100
OT	—	—	—	—	20	20	0
Inventory							
1 Month	60	50	10	—	20	100	100
2 Months	—	10	50	90	100	80	—

Mo.	Source	Month of shipment 6	7	8	9	10	11	12	Total production Actual	Capacity
5	STP	[5] **60**	[10] **10**	[15] —	[20] —	—	—	—	**70**	100
5	OT	—	—	—	—	—	—	—	—	20
6	STP	[0] —	[5] **50**	[10] **50**	[15] —	[20] —	—	—	**100**	100
6	OT	[10] —	—	—	—	—	—	—	—	20
7	STP	X	[0] —	[5] **10**	[10] **90**	[15] —	[20] —	—	**100**	100
7	OT	X	[10] —	—	—	—	—	—	—	20
8	STP	X	X	[0] —	[5] —	[10] **100**	[15] —	[20] —	**100**	100
8	OT	X	X	[10] —	—	—	—	—	—	20
9	STP	X	X	X	[0] —	[5] **20**	[10] **80**	[15] —	**100**	100
9	OT	X	X	X	[10] —	—	—	—	—	20
10	STP	X	X	X	X	[0] —	[5] **100**	[10] —	**100**	100
10	OT	X	X	X	X	[10] **20**	—	—	**20**	20
11	STP	X	X	X	X	X	[0] —	[5] **100**	**100**	100
11	OT	X	X	X	X	X	[10] **20**	—	**20**	20
12	STP	X	X	X	X	X	X	[0] **100**	**100**	100
12	OT	X	X	X	X	X	X	[10] **20**	**20**	20
Shipments		60	60	60	90	140	200	220	**830** / 830	

STP — Straight time production   OT — Overtime production

FIGURE 1   Matrix Relates Month of Production to Month of Shipping

Several refinements may be made to this analysis. For example, the cost savings from a projected capacity increase may be computed simply by adjusting the extreme right column of Figure 1 and reworking the analysis.

assign machines to
part schedules

Matrices are extremely useful in assigning machines to meet a parts schedule. Consider a job shop that has been producing three parts against a twelve-month order for a nearby appliance maker. After two months of uneventful operation, the company is requested to take on two more parts to meet an emergency caused by a strike of another vendor's operations. Since it stands a good chance of retaining this business, it would like to accommodate the customer, and so proceeds to estimate the job using the data in Table III.

Machine I is regularly assigned to this type of work and has a capacity of only 120 hours per week. Hence it is necessary to use two less-efficient machines in taking on the new parts. The characteristics of the alternate machines compared with the original are shown in Table IV.

Total variable cost for each part as produced on each machine is estimated as shown in Table V. Machines II and III will not produce all parts satisfactorily (shown by blank spaces in the table). A production plan is needed which will meet the new schedule at minimum cost. This can be done with the basic matrix procedure only if both the capacity of all machines and the output of all parts can be measured in the same units. For this reason the capacity of each machine is expressed in terms of equivalent hours used by Machine I, as shown in Table VI.

The total adjusted capacity of 216.7 hours per week is composed of exactly equivalent hours from each machine. These hours may therefore be scheduled on the various machines without further concern for their original differences in efficiency.

TABLE III.   THREE PARTS ALREADY MADE BY SHOP
AND TWO ADDITIONAL PARTS THAT ARE BEING CONSIDERED

	Present			*New*	
Part	P-1	P-2	P-3	P-4	P-5
Output rate (pcs/hour)	10	20	7	6	18
Schedule (pcs/wk)	250	480	340	300	600

Appendix

TABLE IV.   MACHINES II AND III, WHICH MUST BE CALLED INTO PRODUCTION, HAVE THESE CHARACTERISTICS

Machine:	I	II	III
Relative efficiency	100%	80%	60%
Hours/wk available	120	80	80
Downtime	5%	7%	10%
Crew size	1.00	1.20	1.20

TABLE V.   TOTAL VARIABLE COST OF EACH PART PRODUCED ON EACH OF THREE MACHINES

Part:		P-1	P-2	P-3	P-4	P-5
Machine—	I	1.75	0.90	2.53	3.19	0.94
	II	1.98	1.02	2.70	–	1.15
	III	–	0.96	–	3.35	0.98

The parts schedule itself is now expressed in terms of Machine I hours simply by dividing the weekly scheduled number of pieces by Machine I's hourly production rate, as shown in Table VI.

TABLE VI.   CAPACITY OF EACH MACHINE EXPRESSED IN TERMS OF EQUIVALENT HOURS USED BY MACHINE I

Machine	Gross	Hours per week			Relative effic.	Adjusted capacity
		Down%	Down hrs.	Net		
I	120	5	6.0	114.0	100%	114.0
II	80	7	5.6	74.4	80	59.5
III	80	10	8.0	72.0	60	43.2
Total	280		19.6	260.4		216.7

(continued)

Parts schedule

Part:	P-1	P-2	P-3	P-4	P-5	Total
Weekly hours:	25.0	24.0	48.6	50.0	33.4	181.0

Equivalent machine I hours						Production	
Machine	P-1	P-2	P-3	P-4	P-5	Plan	Capacity
I	0.85	0	1.63	2.29	0.04	114.0	114.0
	25.0	—	39.0	50.0	—		
II	1.08	0.12	1.80		0.25	23.8	59.5
	—	14.2	9.6	XX	—		
III		0.06		2.45	0.08	43.2	43.2
	XX	9.8	XX	—	33.4		
Demand	25.0	24.0	48.6	50.0	33.4	181.0	216.7

FIGURE 2   Five Parts Are Scheduled on Three Machines of Varying Capacities

The total of 181 hours represents the weekly demands being made on the adjusted capacity of 216.7 hours per week. This total, and its breakdown by parts, is completely compatible with the machine capacity computed earlier. Thus a matrix can be established and unit cost data used to find the lowest cost production plan, Figure 2. As before, the unit cost data is expressed in terms of its differences over the lowest cost time; in this case, $0.90.

The solution in Figure 2 is easily translated into a parts schedule by multiplying the equivalent hours assigned to each part by Machine I's output rate on that part. The matrix may also be turned into a man-hour budget by dividing Machine I equivalent hours by machine efficiency and again by the activity ratio (100 minus downtime percent).

LINEAR

PROGRAMMING TO

AID RESOURCE

ALLOCATION IN

R&D*

Abstract: A capital budgeting problem is formulated and solved by using linear and integer programming. Care is taken to describe fully how constraint equations are developed for various resource requirements.

discussion

The R & D manager in industry is faced with a complex set of resource allocation problems. Some of the most important of these are:

1. how much money to allocate to the R & D area?

2. how much fundamental research not directly aimed at economic benefits should be undertaken?

3. which projects should be undertaken or continued, and which rejected or terminated?

4. at what rates and timings should resources be supplied to each of the selected projects?

Easy answers to questions of this type are rarely possible, but several management techniques now under development may prove of great assistance. The objective of this article is to present one of these techniques—resource allocation by linear programming—which has potential as an aid to resource planning in the short and medium term. It is hoped that the presentation will encourage R & D managers to test the feasibility and desirability of this method, and by so doing provide a feedback to management scientists of its short-comings and deficiencies.

Linear programming is a method which allows a search to be made amongst alternative ways of allocating resources to a set of projects, where limited resources preclude many of the otherwise feasible allocations. The

*By A. G. Lockett and A. E. Gear. Reprinted by permission of the publisher from *The Journal of Management Studies*, May 1970 (Vol. 7, No. 2), pp. 172-182.

outcome of the search is the optimal way of allocating resources (so far as the input data is concerned) in order to maximize the resulting benefits to the organization. Examples of applications of linear programming in a number of fields are described in a book by Vajda.[1]

Weingartner[2] has discussed the application of mathematical programming to capital budgeting problems, and Bell *et al*[3] and Beattie[4] have described linear and integer-based models for application in R & D. The present article is an attempt to describe these latter models in a readily usable form.

The main problem connected with the practical application of linear programming to resource allocation in R & D is the generation of the data requirements. Essentially, these requirements are:

1. a division of the resources into distinct types with the availability of each defined (*resource analysis*);

2. estimation of the resource requirements of each of the project opportunities through time (*project analysis*);

3. evaluation of the eventual expected net benefit of a particular project opportunity if undertaken (*outcome analysis*).

These requirements are described as a series of steps in the following sections, and then the structure of the linear programming model is illustrated with a small data example.

steps

The essential steps involved are listed and described below in what approximates to a logical order.

(a) *Resource Analysis*

　　a1. The resources which are liable to be in short supply are categorized by type. Manpower resources should be categorized in terms of

---

[1] Vajda, S., *Readings in Mathematical Programming*, London: Pitman's Press, 1958.

[2] Weingartner, H. M., *Mathematical Programming and the Analysis of Capital Budgeting Problems*, Chicago: Markham Publishing Co., 1967.

[3] Bell, D.C., Chilcott, J. E, Read, A. W., and Salway, R. A., 'Application of a Research Project Selection Method in the North Eastern Region Scientific Services Department,' *C.E.G.B., R & D Department Report No. RD/H/R₂*.

[4] Beattie, C.J., 'Applications of Mathematical Programming Techniques,' Paper Presented at the *Proceedings of The NATO Conference*, English Universities Press, June 1968.

specialized knowledge, technical skills, supervisory skill, etc. Other resource types may include certain testing and workshop facilities; budgetary monies for purchasing equipment or for running pilot plant, etc.

a2. An overall resource planning period is selected, and broken down into one or more sub-periods depending on the degree of detailed control required. For example, the overall period could be two years from 'time now', broken down into two one-year sub-periods.

a3. The overall availability of each defined resource type is evaluated in each of the sub-periods. Depending on the resource type, availabilities may be defined, for example, in units of man-months, £s, weeks, etc. The availabilities should be net of all allowances for holidays, sickness, free time, service, or regularly occurring work of a mandatory nature, etc.

(b) *Project Analysis*

b4. All the projects, both on-going and new ones, to which the available resources could be allocated, are listed. Any mandatory projects are noted on this list.

b5. Alternative versions of each project, if any, are defined. These may, for example, arise from:

(i) alternative rates of progressing a project. That is, from different rates of resource consumption;

(ii) alternative starting times of a project. For example, the start could be delayed until a later sub-period than the first.

(iii) alternative technical approaches to a project. For example, if two technical methods are being tried out in some application, a series or parallel strategy might be considered.

b6. Estimates are made of the types and amounts of resources which alternative versions of each project would require if undertaken. The amounts of resource are defined in the same units as the corresponding availabilities. Assistance in arriving at these estimates may be provided by network-based planning techniques.

(c) Outcome Analysis

c7. The foreseeable alternative technical outcomes of each project version, if undertaken, are listed. In some cases only one outcome, success, may be foreseen; in others success or failure may need to be considered. In general, a range of technical outcomes may be possible.

c8. Subjectively (or sometimes analytically) based estimates are made of the probability of achieving a given technical outcome of a particular project version, assuming that the version is undertaken with the resources estimated in Step 6.

c9. Assuming a given technical outcome of a particular project version is achieved, an estimate of the resulting economic benefit to the organization is made. The estimate should be net of only those costs which result from the decision to undertake the project. The economic benefit may be calculated as an expected 'present value'. That is all estimated revenues and outlays up to a 'benefit horizon' connected with the project are discounted back to a common point in time using an appropriate discounting rate. In complex situations involving a series of future decisions and/or uncertainties, the estimation of an expected benefit may be aided by constructing and analysing a branching tree diagram. This approach is described by Hespos *et al.*[5]

c10. An overall expected benefit for each project version is calculated, assuming the version is undertaken. This is done by multiplying together the expected benefit of each technical outcome from Step 9 with its estimated probability of achievement from Step 8 and then adding together all the products. The resultant number is the overall expected benefit of the project version, based on the data estimates.

model structure

The foregoing steps should provide the essential information to enter a linear programming model: definition of resource types and availabilities; estimation of the resource requirements of each project version; a value for the expected benefit of each project version if undertaken. The model, which would normally be run on a computer as a considerable amount of arithmetic is involved, can now be solved to determine that allocation of resources within the overall R & D planning period which will maximize the total expected benefit deriving from the projects between time now and the benefit horizon.

Example:

In Table I the collected data requirements of a fictitious four project example are presented. The projects, numbered 1 to 4, are planned in 1, 4, 2, and 1

[5] Hespos, R. F., and Strassman, P.A., 'Stochastic Decision Trees for the Analysis of Investment Decisions,' *Management Science*, Vol. II, No. 10, August 1965.

TABLE I

			Sub-Period I				Sub-Period 2				Expected benefit of given project version
			Phys	Chem I	Chem II	Budget	Phys	Chem I	Chem II	Budget	
Project 1	Version 1	x_{11}	10	10	25	2	5	5		2	1.8
Project 2	Version 1	x_{21}	15	13	10	5	8	5	10	8	6.0
	2	x_{22}	8	5		5	15	13	25	5	4.8
	3	x_{23}					8	5	10	5	5.0
	4	x_{24}					3		7	2	3.5
Project 3	Version 1	x_{31}	3		7	2					8.5
	2	x_{32}	6		20	2					10.0
Project 4	Version 1	x_{41}		25	10	10		25	10	10	13.0

technical versions respectively. In the third column of the table a variable x_{ij} is assigned to each project version. Suffix i represents the project number, while suffix j represents the version number. Thus x_{23} is the variable assigned to project 2, version 3.

The variables x_{ij} are introduced so that the problem can be formulated as a mathematical programming problem. If in the solution a variable takes a value of unity, then this implies that the particular project version should be undertaken as planned. On the other hand, a value of zero implies that the given project version should be rejected.

In the table it will be seen that estimates of the resource requirements of each project version are given for each of four types of resource and in two forward sub-periods of time. Three of the resource types are manpower (physicists, chemists of type I and II), while the fourth type is monetary. Thus, for example, version 2 of project 3 requires: 6 units of physics manpower (e.g. 6 man-months); 20 units of chemistry II manpower; 2 units of a budget money (e.g. £$2 \cdot 10^3$), all in period I.

The final column of the table is the expected benefit of each of the project versions in suitable units of money (e.g. £10^5).

In Table II the estimated overall availabilities of each of the four resource types in the two sub-periods are listed.

In Table III the data from Tables I and II is presented in the form of a mathematical programming matrix. Each of the columns is headed by a variable x_{ij} and it is implied that each of the numbers appearing in a particular column is multiplied by the column heading variable. The rows are identified by a row number and a description, and each row represents a different constraint which is imposed on the situation. The last two columns indicate the sign of each constraint row and the limiting value of the constraint. The final row is the objective function to be maximized.

Constraint Rows.   The rows numbered 1 to 4 in Table III are provided to ensure that each of the four projects is not selected more than once. It will be noticed that row 3 carries an equals (rather than a less than or equals) sign. This further ensures that project 3 will definitely be fully selected in some version, thus making project 3 mandatory. Written out in full, row 2 becomes:

$$x_{21} + x_{22} + x_{23} + x_{24} \leqslant 1$$

and similarly for rows 1, 3, and 4.

TABLE II

		Availability
Sub-Period 1	Phys.	20
	Chem. I	40
	Chem. II	30
	Budget	21
Sub-Period 2	Phys.	20
	Chem. I	40
	Chem. II	30
	Budget	21

An interesting feature of the model is that two alternative viewpoints regarding the values which the variables x_{ij} are allowed to take may be adopted. These are:

A. All the x_{ij} variables can take any value between 0 and 1. In this case a linear programming computer package is used;

or

B. All the x_{ij} variables can take only the values 0 or 1. In this case an integer (or mixed integer) programming computer package is required.

These alternatives are discussed further in the section headed *outputs*.

The rows numbered 5 to 12 are provided to ensure that the total amount of each type of resource required in the selected set of projects does not exceed its respective overall resource availability. For example, row 5 ensures that the availability of 20 units of physics manpower is not exceeded in sub-period 1. Written out in full this row is:

$$10x_{11} + 15x_{21} + 8x_{22} + 3x_{31} + 6x_{32} \leqslant 20$$

Row 17 is the objective function to be maximized. That is, the objective of the resource allocation is to find the values of the variables x_{ij} such that the value of the objective function

$$1.8x_{11} + 6.0x_{21} + 4.8x_{22} + 5.0x_{23} + 3.5x_{24} + 8.5x_{31} + 10.0x_{32} + 13.0x_{41}$$

is a maximum subject to the constraints.

TABLE III

Row No.	Row Description	x_{11}	x_{21}	x_{22}	x_{23}	x_{24}	x_{31}	x_{32}	x_{41}	Sign	R.H.S.
1	Project 1	1								≤	1
2	Project 2		1	1	1	1				≤	1
3	Project 3						1	1		=	1
4	Project 4								1	≤	1
5	Phys. Per. 1	10	15	8			3	6	25	≤	20
6	Chem. I. Per. 1	10	13	5					25	≤	40
7	Chem. II. Per. 1		25	10			7	20	10	≤	30
8	Budg. Per. 1	2	5	5			2	2	10	≤	21
9	Phys. Per. 2	5		8	15	8	3			≤	20
10	Chem. I. Per. 2	5		5	13	5			25	≤	40
11	Chem. II. Per. 2			10	25	10	7		10	≤	30
12	Budg. Per. 2	2		8	5	5	2		10	≤	21
17	Objective (Max.)	1.8	6.0	4.8	5.0	3.5	8.5	10.0	13.0		

Outputs.   Two alternative solutions to the example problem of Table III are presented in Table IV. These are labelled A and B, and are the optimal solutions resulting from processing the data with a linear and integer programming algorithm respectively.

Solution B is the easiest to interpret, indicating the particular version of each project which is selected. The solution chooses versions 1 of projects 1 and 4. This means that these projects should be done in the first period, i.e. now. It also chooses version 4 of project 2 and version 2 of project 3, i.e. project 3 should be started now, but project 2 should not be started until period 2. This solution happens to contain a version of all the four projects, so that the solution is indicating a near-optimal schedule for allocating resources rather than just indicating selection. (It could have been that one or more of the projects was not selected in any version.)

Solution A agrees with Solution B for projects 1, 3 and 4, but differs from it for project 2. Taken literally, Solution A for project 2 means that 0.27 of the resources of version 1, plus 0.40 of the resources of version 3, plus 0.33 of the resources of version 4 should be the optimal pattern of resource usage. This derived version of project 2 is assumed to have a benefit of $0.27 \times 6.0 + 0.40 \times 5.0 + 0.33 \times 3.5$, which is the sum of the benefits of the versions of project 2 weighted down by the respective fractions selected. If this linear assumption is acceptable, and if the derived resource allocation pattern for the project is feasible, then the Solution A may be preferred to B in view of the slightly greater value of the objective function. In both cases the utilization of manpower for the solutions is of the order of 90 per cent, which is satisfactorily high.

In this example the sum of the fractions of versions of project 2 in Solution A totals to unity. If this total is fractional the same type of investigation of acceptability and feasibility is undertaken, but there is less chance that the above described assumption will be acceptable. Further practical experience is required before the relative merits of these approaches, or of some composite approach, can be evaluated.

Note that rounding off, *i.e.* choosing version 3 of project 2, does not produce the same results as the integer programming solution which chooses version 4 of project 2. Furthermore this approach is liable to lead to an infeasible solution: in this case the constraints for Chemists Class II are broken in both periods and for Class I Chemists in period 2.

model extensions

The basic model described above is capable of including a number of additional features in order more closely to simulate a given situation. In particular, flexibility of certain skilled manpower to work in more than one manpower

TABLE IV

Project	Version	Variable	Solution A(LP)	Solution B(IP)
1	1	x_{11}	1.0	1.0
2	1	x_{21}	0.27	0.0
2	2	x_{22}	0.0	0.0
2	3	x_{23}	0.40	0.0
2	4	x_{24}	0.33	1.0
3	1	x_{31}	0.0	0.0
3	2	x_{32}	1.0	1.0
4	1	x_{41}	1.0	1.0

Value of objective function	= 29.6	28.3
Utilization of period 1 manpower	= 90.5%	90%

category can be included. This extension is fully described by Bell *et al*[6] and is only briefly presented in what follows.

In terms of the example introduced earlier, suppose that up to 10 per cent of physicists can undertake the work of chemists of type I, and that up to 25 per cent of chemists of type I can do the work of chemists of type II. This laboratory situation can be diagrammatically represented as shown below:

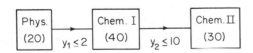

The availabilities in the absence of manpower transfer are the bracketed numbers. The arrowed areas between the resource categories indicate the transfer directions. A variable is introduced to represent the amount of each transfer. The only constraint on these variables is an upper limit on the value of each in keeping with the actual laboratory situation.

The shape of the extended matrix of Table III to include the exampled transfers is shown in Table V. It is necessary to allow for the transfer possibilities in the two periods, and so the variables y_1 and y_2 are used for sub-period 1 and the variables y_3 and y_4 for sub-period 2. The ($+1$) and (-1) entries in the 'y' columns arise because the availabilities of physicists, chemists I and chemists II are modified to $20 - y_1$, $40 - y_1 - y_2$ and $30 + y_2$ respectively because of the

---

[6] Bell, D. C., Chilcott, J.E., Read, A.W., and Salway, R.A., op. cit.

TABLE V

Row No.	Row Description	x_{11}	x_{21}	x_{22}	x_{23}	x_{24}	x_{31}	x_{32}	x_{41}	y_1	y_2	y_3	y_4	Sign	R.H.S.
1	Project 1	1												≤	1
2	Project 2		1	1										≤	1
3	Project 3				1	1	1	1						=	1
4	Project 4								1					≤	1
5	Phys. Per. 1	10	15	8			3	6		+1				≤	20
6	Chem. I. Per. 1	10	13	5					25	−1	+1			≤	40
7	Chem. II. Per. 1		25	10			7	20	10		−1			≤	30
8	Budg. Per. 1	2	5	5			2	2	10					≤	21
9	Phys. Per. 2	5		8	15	8	3					+1		≤	20
10	Chem. I. Per. 2	5		5	13	5			25			−1	+1	≤	40
11	Chem. II. Per. 2			10	25	10	7		10				−1	≤	30
12	Budg. Per. 2	2		8	5	5	2		10					≤	21
13	Transfer 1									1				≤	2
14	Transfer 2										1			≤	10
15	Transfer 3											1		≤	2
16	Transfer 4												1	≤	10
17	Objective (Max.)	1.8	6.0	4.8	5.0	3.5	8.5	10.9	13.0						

transfer possibilities. The additional four rows, numbered 13 to 16 inclusive, place upper limits on the transfer variables of 2 for y_1 and y_3 and 10 for y_2 and y_4.

Other features which can be added to the model include: an allowance for future opportunities which are likely to arise during the overall planning period; the addition of a constraint row which places a limit on the overall variability of total benefits of the selected programme.

Difficulties.   A general difficulty associated with the technique is its detailed data requirements, which necessitate the establishment of a management information system, so that the technical and commercial data can be integrated.

The model requires an estimate of the expected value of each version of those projects which are not mandatory. This estimate may be obtained by techniques based on risk analysis. The assumption which is built in to the model is that a *definite* amount of resources are consumed by a particular project version *before* one of a number of possible technical outcomes is achieved.

While this assumption is applicable to many projects, it may be erroneous for others. It is sometimes the case, for example, that an unsatisfactory outcome can be foreseen well in advance of consuming all the resources originally planned to achieve a satisfactory outcome. This is an example of the failure of the model as described to allow for the sequential learning process involved in R & D.

conclusions

Mathematical programming based on linear and/or integer programming models may have an important part to play in aiding R & D management with respect to resource allocation decisions in the short and medium term. The models require the R & D resource situation to be analysed and the opportunities for R & D resource situation to be analysed and the opportunities for R & D projects to each be planned and evaluated.

The present models cater for a 'deterministic' resource consumption for a given opportunity followed by a stochastic return. An assumption is implied that one of a foreseen set of technical outcomes is only achieved if and after the planned set of resources are consumed. This may introduce an error, especially where the project strategy is arranged to investigate the most serious technical problem early on in the project's life in order to reassess the value of continuation.

INDEX

Index

U

V